Topics in
Current Physics

9

# Topics in Current Physics

Founded by Helmut K.V. Lotsch

Volume 1 **Beam-Foil Spectroscopy**
Editor: S. Bashkin

Volume 2 **Modern Three-Hadron Physics**
Editor: A. W. Thomas

Volume 3 **Dynamics of Solids and Liquids by Neutron Scattering**
Editors: S. W. Lovesey and T. Springer

Volume 4 **Electron Spectroscopy for Surface Analysis**
Editor: H. Ibach

Volume 5 **Structure and Collisions of Ions and Atoms**
Editor: I. A. Sellin

Volume 6 **Neutron Diffraction**
Editor: H. Dachs

Volume 7 **Monte Carlo Methods** in Statistical Physics
Editor: K. Binder

Volume 8 **Ocean Acoustics**
Editor: J. DeSanto

Volume 9 **Inverse Source Problems** in Optics
Editor: H. P. Baltes

Volume 10 **Synchrotron Radiation,** Techniques and Applications
Editor: C. Kunz

Volume 11 **Raman Spectroscopy** of Gases and Liquids
Editor: A. Weber

Volume 12 **Positrons in Solids**
Editor: P. Hautojärvi

Volume 13 **Computer Processing of Electron Microscope Image**
Editor: P. W. Hawkes

Volume 14 **Excitons**
Editor: K. Cho

Volume 15 **Physics of Superionic Conductors**
Editor: M. B. Salamon

# Inverse Source Problems

## in Optics

Edited by H. P. Baltes

With Contributions by
H. P. Baltes   H. A. Ferwerda   J. Geist
B. J. Hoenders   H. G. Schmidt-Weinmar
A. Walther   A. Zardecki

With a Foreword by J.-F. Moser

With 32 Figures

Springer-Verlag Berlin Heidelberg New York 1978

Professor Dr. Heinrich P. Baltes
LGZ Landis & Gyr Zug AG, Zentrale Forschung und Entwicklung
CH-6301 Zug, Switzerland



ISBN 3-540-09021-5 Springer-Verlag Berlin Heidelberg New York
ISBN 0-387-09021-5 Springer-Verlag New York Heidelberg Berlin

Library of Congress Cataloging in Publication Data. Main entry under title: Inverse source problems in optics. (Topics in current physics ; v. 9) Bibliography: p. Includes index. 1. Optics, Physical. 2. Inverse problems (Differential equations) I. Baltes, Heinrich P. II. Series. QC395.2.I58 535'.2 78-12076



Offset printing and bookbinding: Konrad Triltsch, Graphischer Betrieb, Würzburg.
2153/3130-543210

# Foreword

*Puis, lorsque j'ai voulu descendre à celles (les choses) qui étaient plus particulières, il s'en est tant présenté à moi de diverses, que je n'ai pas cru qu'il fût possible à l'esprit humain de distinguer les formes ou espèces de corps qui sont sur la terre, d'une infinité d'autres qui pourraient y être si c'eût été le vouloir de Dieu de les y mettre, ni par conséquent de les rapporter à notre usage, si ce n'est qu'on vienne au-devant des causes par les effets, et qu'on se serve de plusieurs expériences particulières.* [*R. Descartes: Discours de la méthode (Librairie Ch. Delagrave, Paris 1877) Part 6, p.65*]

It is an interesting fact that text-book physics is characterized by a strong predilection for direct problems, that is predicting physical effects on the basis of known physical causes. The complex mathematical apparatus involved in solving inverse problems, especially the type covered in this book, presents serious difficulties to the beginner in the field. He might also be surprised that a purely industrial need is at least partially responsible for the writing of this book, a hint that the field of inverse problems in optical physics has grown beyond mathematical art.

What is it then which makes, for example, the apparatus industry take a close look at the inverse problem in optical physics? Let me give one example from the market of high- and low-speed banknote testing equipment. The function of such machines is to test a graphic product, the banknote, for its genuineness with the highest degree of security and reliability. The introductory chapter gives a direct link to the type of problem encountered in optical authenticity checking.

However, this book will not reveal practical approaches to the solution of technical problems. The road leading to the technical implementation of the results achieved so far is long and not easy.

Chapter 1 attempts a brief systematic survey of the inverse problems in optical physics, together with a discussion of the role of prior knowledge. The chapter presents a tentative list of more than 20 specific inverse optical problems (including those not covered by this volume). The agreed size of the Topics in Current Physics series volumes imposed the selection of not more than the following five chapters.

Chapter 2 presents a state-of-the-art review of the phase reconstruction problem for wave amplitudes, as well as coherence functions with application to both light and electron optics.

Chapter 3 is devoted to the problem of reconstructing a scattering object or potential from scattered field amplitudes with emphasis on the question of uniqueness and nonradiating sources. The reconstruction of the field up to the surface of the scatterer, as well as the reconstruction of the object from the field outside the object, are described in detail.

Chapter 4 reports recent work toward solving a superresolution problem, namely the reconstruction of the near field of very small localized sources from far-field data. As a by-product, this chapter contains a comprehensive study of nonuniform plane waves.

Chapter 5 aims at the new field where coherence and radiometry overlap. The relationship between far-zone and source coherence functions is discussed along with new radiometric concepts for sources of any state of coherence. Both amplitude and intensity correlations are considered. A brief survey of the history of radiometry is also given.

Finally, Chapter 6 reviews the retrieval of statistical features of random phase screens from scattering data in terms of correlation functions and photon statistics. Higher-order statistical properties of the scattered field are emphasized. Nonchaotic scattered radiation due to a small number of scatterers is discussed.

We believe this is the first book written on the subject; too few people have been taken this road. May this book invite others to join in the effort to widen the potentials of this particular field.

Zug, Switzerland
July 1978 *J.-F. Moser*

# Contents

# List of Contributors

BALTES, HEINRICH P.

Zentrale Forschung und Entwicklung, Landis & Gyr Zug,
CH - 6301 Zug, Switzerland

FERWERDA, HEDZER A.

Vakgroep Technische Fysica, Rijksuniversiteit Groningen,
Nijenborgh 18, 9747 AG Groningen, The Netherlands

GEIST, JON

Radiometric Physics Division, Center for Radiation Research,
National Bureau of Standards, Washington DC 20234, USA

HOENDERS, BERNHARD J.

Vakgroep Technische Fysica, Rijksuniversiteit Groningen,
Nijenborgh 18, 9747 AG Groningen, The Netherlands

SCHMIDT-WEINMAR, HEINZ GÜNTER

Department of Electrical Engineering, University of Alberta,
Edmonton, Alberta, Canada TG6 2E1

WALTHER, ADRIAAN

Physics Department, Worcester Polytechnic Institute,
Worcester MA 01609, USA

ZARDECKI, ANDRZEJ

Laboratoire de Recherche en Optique et Laser, Département de Physique,
Université Laval, Québec, Canada G1K 7P4

# 1. Introduction

## H. P. Baltes

We begin the introductory chapter with a general definition of the inverse optical problem. Next, we discuss the role of prior knowledge and the questions of uniqueness and stability. We then review the various specific inverse problems in optics as well as the contents of Chapters 2 to 6. Finally, we summarize the notation in coherence theory.

## 1.1 Direct and Inverse Problems in Optical Physics

The "direct" or "normal" problem in optical physics is to predict the emission or propagation of radiation on the basis of a known constitution of sources or scatterers. The "inverse" or "indirect" problem is to deduce features of sources or scatterers from the detection of radiation. An intuitive solution of the optical inverse problem is commonplace: we infer the size, shape, surface texture, and material of objects from their scattering and absorption of light as detected by our eyes. Intuition has to give way to mathematical reconstruction as soon as we wish to analyze optical data beyond their visual appearance. Examples are the extrapolation and deblurring of optical images, the reconstruction from intuitively inaccessible data such as defocused images and interferograms, or the search for information that is "lost" in the detection process such as the phase.

Following CHADAN and SABATIER [1.1], a general definition of inverse optical problems can be attempted as follows. We describe the sources and scatterers by the set

$$G = \{g_1, g_2, \ldots, g_n\} \qquad (1.1)$$

of space-time functions $g_j$ which we call the *source functions* (scatterers being included as indirect or secondary sources). The resulting propagation of radiation is described by the set

$$F = \{f_1, f_2, \ldots, f_m\} \qquad (1.2)$$

of space-time functions $f_i$ called *results* or *data*, which can be checked by measurement. From the source functions $g_j$, we can derive unique data $f_i$ by virtue of the *direct relations*

$$f_i = E_i(g_1, g_2, \ldots, g_n) \quad , \tag{1.3}$$

where the set $E$ of operators $E_i$ provides a mapping of $G$ into $F$, viz.

$$E : G \to F \quad . \tag{1.4}$$

In coherent optics, for example, the $E_i$ correspond to certain integral transformations and the $g_j$ and $f_i$ to source and, for example, far-zone amplitudes and their correlations.

*Solving the direct problem* in optical physics means computing the data $f_i$ from known source functions $g_j$ using the direct relations (1.3). *Solving the inverse problem* means finding source functions $g_j$ which

1) correspond to given data $f_i$ by virtue of the prescriptions (1.3) and

2) are consistent with the physical information coming from general principles or other experiments, the so-called *prior knowledge*.

The prior knowledge reduces the set of possible source functions. For example, we can often take for granted that the source has a finite volume.

Apparently there are two opposite approaches to the above problem.

1) We establish formulas or algorithms which allow the reconstruction of the source functions by inversion of the mapping (1.3,4), viz.

$$E^{-1} : F \to G \quad . \tag{1.5}$$

The name "inverse problem" is usually reserved for this approach.

2) We find specific model source functions by trial and error and fit free parameters from the experimental data. This approach brings us back to the direct problem, since we have to check the models by (1.3). The notion of inverse problem in the strict sense is usually understood to exclude such fitting procedures.
In practice there is, however, a more or less continuous transition from "inverse" to "direct" procedures, the inverse character of the problem becoming less pronounced with increasing prior knowledge (see Sect.1.2).

It is well known that an inversion of mappings as indicated by (1.5) involves the mathematical questions of the *existence*, *uniqueness*, and *stability* of the solution. For example, the extrapolation of optical image data [1.2,3] belongs to the class of problems (usually called "ill-posed" or "improperly posed problems"), in which the solution depends uniquely, but not continuously, on the data (see, e.g., [1.4]). Small errors in the data can lead to large errors in the solution unless suitable stabilizing constraints are imposed, i.e., unless additional prior knowledge can be taken for granted. Of course, errors and noise are inevitable in experimental data.

As for historical remarks, we refer to the introduction of Chapter 3 (see also [1.1]).

## 1.2 Role of Prior Knowledge

Let us now attempt to re-collect the prerequisites for obtaining information on an optical source or a scatterer (or the propagating medium) from experimental data. The following scheme may be helpful.

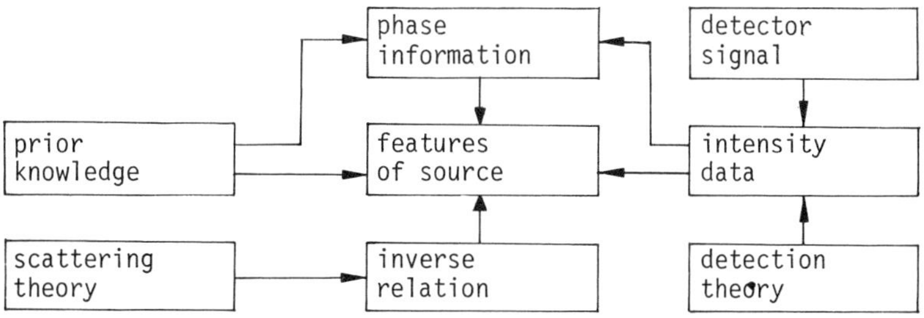


The sought features of the source or the scatterer are, in principle, inferred from intensity and phase data by virtue of the appropriate inverse relation, and accounting for the available prior knowledge. The inverse relation or inversion algorithm is, of course, based on the pertinent theory of propagation and scattering of radiation. Detectors provide intensity data. We thus have to deduce the necessary phase information from intensity distributions. This problem of *phase reconstruction* is the objective of Chapter 2. Intensity data can also include quantities from coherent and quantum optics such as the modulus of the degree of coherence, the intensity autocorrelation, and photon statistics. The correct evaluation of detector signals requires knowledge of the *theory of photodetection* [1.5] and involves another inverse problem, namely how to reconstruct the statistics of the incident radiation from that of the photoelectrons [1.6].

*Prior knowledge* means simply any knowledge about the source functions available prior to the experiment in which we are interested, but, of course, not prior to the development of the plan of observation [1.7]. Such knowledge is inferred from general principles, hypotheses [1.8], the result of other experiments, and the constraints imposed by the planned experimental procedure. The notion of prior knowledge used here is distinct from epistemological or a priori knowledge in the strict Kantian sense (see [1.7]). Prior knowledge is crucial for achieving uniqueness and stability of the solution of the inverse problem. Moreover, the nature of the prior knowledge largely determines the character of the problem (inverse or direct). If sufficient prior knowledge allows us to infer specific source models, we get away with solving the direct problem and fitting the parameters, as is illustrated by the following example.

1) We begin the well-known determination of stellar diameters from measurements of the modulus $|\mu|$ of the degree of coherence $\mu$ as a function of angular spacing (see, e.g., [Ref.1.2, Sect.2]). An enormous amount of prior knowledge is taken for granted here: we assume that the source is a uniformly bright, circular disk with zero coherence area. Applying the Van Cittert-Zernike theorem to this model, we learn how the angular diameter of the source is found from the first zero of $|\mu|$. We had, therefore, to solve a "weak" inverse problem, i.e., to establish and evaluate nothing but a direct relation.

2) Let us now drop the prior knowledge on the shape of the source. By the Van Cittert-Zernike theorem, $\mu$ is the Fourier transform of the intensity profile $I_0$ in the source plane. Thus the shape and size of the source can, in principle, be determined by inversion of the Fourier transform relationship. However, we now have to measure $|\mu|$ over a large range, and must possibly reconstruct the phase of $\mu$. Moreover, we are faced with a serious extrapolation problem if we ask for small details of the source intensity distribution.

3) If neither the intensity distribution $I_0$ nor the degree of spatial coherence $\mu_0$ in the source plane is known, we have a still more complicated inverse problem involving the convolution of $\mu_0$ with the autocorrelation of $I_0^{1/2}$ (see Sect.5.4.2). Without further data (e.g., the radiant intensity), the measurement of $|\mu|$ is not sufficient for disentangling the information on $I_0$ and $\mu_0$.

Concluding this section we emphasize that the consideration of stability questions and the exact specification of the prior knowledge are indispensable.

## 1.3 Survey of Specific Inverse Problems

This volume presents only a small number of selected topics out of the many (20 or more) specific inverse problems of optical physics. In this section we attempt to list the various problems, including those not to be covered in this book (and a few that have hardly been attacked yet). Some readers may be concerned about what is *not* to be found in this book. A selection (five out of ten originally planned chapters) was imposed by the agreed size of the Topics in Current Physics series volumes.

Perhaps a future complementary volume will improve the situation.

In principle, we can distinguish two classes of inverse optical problems, namely

1) problems aiming at information on *spatial* variations of the source functions (spatial frequency spectra), such as the intensity profile or the degree of spatial coherence and other space correlations, and

2) problems aiming at information on *time* variations (dynamics) of the source functions (time frequency spectra) such as the spectral density or the degree of temporal coherence and other time correlations.

In the present volume we consider mainly inverse problems of the type 1). We notice, however, that speckle patterns in polychromatic light [1.9] and scattering by moving diffusers (see, e.g., [1.10]) involve time *and* space variation and are included in Chapter 6. Another inverse problem combining spectral and spatial aspects is the reconstruction of the shape of a cavity resonator from the eigenvalue spectrum (or temporal coherence function) mentioned in Chapter 5.

Another possible classification of inverse problems can be based on the statistical aspect of the radiation. Thus we have inverse problems in classical radiative transfer ("transport of intensity" in the limit of poor spatial coherence), wave optics (coherence limit), and coherent and quantum optics. This volume presents a selection of inverse problems with wave amplitudes (Chaps.3 and 4, first part of Chap.2) and coherence functions (Chaps.5 and 6, second part of Chap.2). Inverse radiative transfer is not studied in this book.

Including related mathematical questions, as well as a number of "applied problems", we arrive at the following, probably incomplete, *list of inverse problems in optics*. (The asterisk indicates that the problem is treated in this volume.)

---

1. Intensity propagation
   - 1.1 Inverse radiative transfer (inverse transport theory)
2. Wave amplitudes
   - 2.1 *Phase reconstruction
   - 2.2 *Inverse diffraction (from surface to surface)
   - 2.3 *Inverse scattering (determination of scattering object or potential)
   - 2.4 *Reconstruction of source fields or scattering objects beyond diffraction limit (superresolution problems)
   - 2.5 Extrapolation of images beyond borders
   - 2.6 Computational reconstruction from holographic data
   - 2.7 Reconstruction of optical cavity from the eigenvalue spectrum
   - 2.8 Inverse problems in ellipsometry
3. Coherence functions
   - 3.1 *Phase reconstruction for spatial coherence functions
   - 3.2 *Phase reconstruction for temporal coherence functions
   - 3.3 *Inversion of radiometric data for planar sources (2D)
   - 3.4 Inverse diffraction and scattering of coherence functions for 3D sources
   - 3.5 Extrapolation and superresolution problems for partially coherent light
4. Statistical states
   - 4.1 Reconstruction of radiation field statistics from detector signals
   - 4.2 Determination of statistical field operators from moments or correlations
   - 4.3 Maximum entropy image restoration (photon statistical aspect)

4.4 Determination of initial statistical state of scatterers (e.g., atoms) from statistical state of scattered radiation

5. General mathematical questions
   5.1 *Uniqueness and nonradiating sources
   5.2 *Stability and constraints

6. Some applied problems
   6.1 Shape (optical density) of particles, fibers, or polymers from scattering data (e.g., Mie scattering)
   6.2 *Features of random phase screens (e.g., rough surfaces) from scattering data
   6.3 Optical communication through, and information on, turbid media (laser range finding, remote sensing, matched optics)
   6.4 Preparation of filters (e.g., phase screens) tailored to predetermined scattering properties

---

Let us briefly comment on the above list. We begin with the *topics referred to this volume. The phase reconstruction problems 2.1 and 3.1 are the objective of Chapter 2. References on the phase problems 3.2 are also given in Chapter 2. Chapter 3 is devoted to inverse diffraction and scattering (problems 2.2 and 2.3) with emphasis on the question of uniqueness (problem 5.1). We mention that problem 2.3 is closely related to the inverse problem in quantum scattering [1.1]. Superresolution for coherent light (problem 2.4) is the topic of Chapter 4. Stability problems are touched upon in Chapters 2 and 4 (see also references given at the end of Sect.3.1.3). In Chapter 5 we consider the radiometry of partially coherent planar sources and the inversion of first-order radiometric data (problem 3.3). The radiometric quantities corresponding to second-order coherence are also considered in Chapter 5, but the pertinent inverse relations have not been established hitherto. Finally, Chapter 6 is devoted to one of the applied problems, viz. how to find statistical features of random screens from optical scattering data. Here, the state of the art is far from allowing a general inversion procedure; therefore specific statistical models cannot be avoided. The logical connections between the different chapters can be seen from the following sketch.

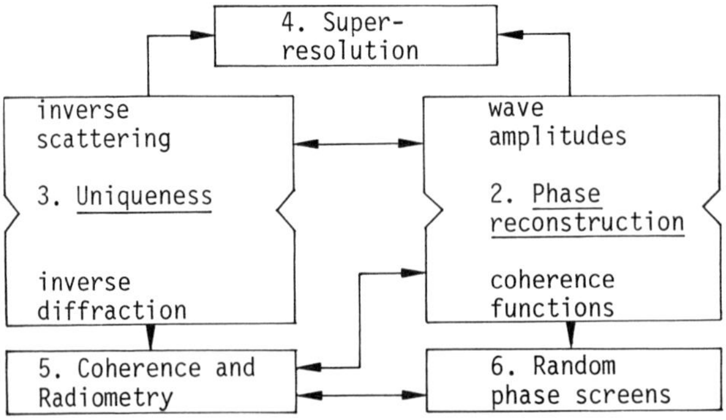

Let us continue with a few references on some of the problems not treated in this volume. Inverse radiative transfer (problem 1.1) is another example of an ill-posed problem (see, e.g., [1.11]). Many references are found in a recent paper by ISHIMARU [1.12], who reviews also the relationship between radiative transfer and the multiple scattering of waves. The extrapolation problem 2.5 is mentioned in [1.3,4]. For the reconstruction from holographic data we refer to [1.13] and references therein. The relationship between shape and mode density of optical cavities is reviewed in a recent book by BALTES and HILF [1.14]; further references are given in Section 5.2.1. An inversion problem of reflection ellipsometry is discussed in [1.15]. Superresolution with partially coherent light in the presence of noise (problems 3.5 and 5.2) is discussed by ROSS [1.7] and PEŘINA et al. [1.33].

The reconstruction of the quasi-probability density from the photoelectric counting rate (problem 4.1) is reviewed in a paper by BERTOLOTTI [1.16]. The photocounting problem is currently reinvestigated by SELLONI et al. [1.17]. Aspects of problem 4.2, namely reconstructions of statistical operators with given moments, were studied by JAYNES [1.18], INGARDEN and KOSSAKOWSKI [1.19] and, most recently, SCHWENDIMANN et al. [1.20]. As for problem 4.3, we mention a recent article by KIKUCHI and SOFFER [1.21] and references therein.

Let us now touch upon some applied problems. There is abundant literature on scattering by small (spherical) particles; see, for example, the references given in [1.12]. Promising correlation studies of light scattered from dense suspensions are due to NARDUCCI et al. [1.22]. The structure of fibers and polymers from scattered radiation has been studied by, for example, ROSS [1.7] and HASHIMOTO et al. [1.23]. As for problem 6.3, we refer to recent reviews by SCHANDA [1.24], HINKLEY [1.25], and SHAPIRO [1.26]. An interesting application of the inverse diffraction problem is the preparation of screens with given scattering properties. First steps toward this objective are computer-generated holograms and kinoforms [1.27]. In addition to the mathematical inversion, a number of technological problems have to be solved here. A new kind of tailored reflection screen with controlled degree of surface randomness and directionality of the scattered light has been realized by ANTES and GREENAWAY [1.28].

To the best of our knowledge, the remaining problems 3.4 and 4.4 have hardly been broached hitherto. In order to solve the problem 3.4, one has to incorporate the machinery of wave-amplitude inverse scattering into coherence theory. Problem 4.4 aims at the inversion of the dynamics of coupled systems in quantum optics.

As for the contact with information and communication theory, we refer to, e.g., [1.3,6-8,18,21].

## 1.4 Notation in Coherence Theory

Let us summarize the optical correlation functions used in Chapters 2, 5, and 6. We start with GLAUBER's *n-th order correlation function* (see, e.g., [1.29])

$$\Gamma^{(n)}(x_1,\ldots,x_n,x_{n+1},\ldots,x_{2n}) = \langle E^{(-)}(x_1)\ldots E^{(-)}(x_n)E^{(+)}(x_{n+1})\ldots E^{(+)}(x_{2n})\rangle \quad (1.6)$$

letting $x_j$ stand for the coordinates of the space-time point $\underline{r}_j$, $t_j$, and $E^{(-)}$, $E^{(+)}$ for the hermitian conjugate electric field operators. (We suppress the vector and tensor notation.) We recall that a radiation field is said to exhibit *n-th order coherence* if and only if all correlation functions of order $m \leqq n$ factorize. The corresponding *degree of coherence* is defined by

$$\gamma^{(n)}(x_1,\ldots,x_{2n}) \equiv \Gamma^{(n)}(x_1,\ldots,x_{2n}) \prod_{j=1}^{2n} \left[\Gamma^{(1)}(x_j,x_j)\right]^{-1/2} \quad . \quad (1.7)$$

The *first-order* correlation function $\Gamma^{(1)}$ is identified with the *mutual coherence function* $\Gamma$ of stationary fields [1.30], viz.

$$\Gamma_{12}(t) \equiv \Gamma(\underline{r}_1,t_1;\underline{r}_2,t_2=t_1+t) = \Gamma^{(1)}(x_1,x_2) \quad . \quad (1.8)$$

The first-order degree of coherence $\gamma^{(1)}$ is likewise associated with the *complex degree of coherence* [1.30]

$$\gamma_{12}(t) \equiv \gamma(\underline{r}_1,t_1;\underline{r}_2,t_2=t_1+t) = \gamma^{(1)}(x_1,x_2) \quad . \quad (1.9)$$

Consideration of a single Fourier component (frequency $\omega$) leads to the "spectral coherence function" or "cross-spectral density"

$$W(\underline{r}_1,\underline{r}_2) = \int_{-\infty}^{+\infty} dt \, \exp(i\omega t)\Gamma_{12}(t) \quad (1.10)$$

and the corresponding "degree of spectral coherence"

$$\mu_{12} = \mu(\underline{r}_1,\underline{r}_2) - W(\underline{r}_1,\underline{r}_2)[W(\underline{r}_1,\underline{r}_1)W(\underline{r}_2,\underline{r}_2)]^{-1/2} \quad . \quad (1.11)$$

In general, the above quantities depend on frequency, but here we dispense with displaying this dependence in the formulae. Of course, these quantities describe the (first-order) *spatial correlation* of the radiation field for the spectral component $\omega$. The notion of "angular correlation" and "angular coherence" or "lateral coherence" is used if $|\underline{r}_1| = |\underline{r}_2|$ and the spatial coordinates (in the far zone) are expressed by angles.

In the strict sense, *full coherence* means $|\gamma^{(n)}| \equiv 1$ for any n, i.e., all $\Gamma^{(n)}$ factorize. Thus *incoherence* indicates that at least one $\Gamma^{(n)}$ does not factorize. In a more relaxed sense, hypothetical sources with $|\mu(\underline{r}_1,\underline{r}_2)| = 0$ unless $|\underline{r}_1-\underline{r}_2|$ is small compared with a wavelength are often called "incoherent" or "fully incoherent", whereas the notion of "full (first-order) coherence" is used for fields with $|\mu(\underline{r}_1,\underline{r}_2)| \equiv 1$.

The *second-order* correlation function $\Gamma^{(2)}$ is associated with the effects of "second-order coherence" in an analogous way.

The above notation is used in the present volume. We notice that some authors (e.g., [1.31,32]) prefer a different notation of the order of coherence and use the name "(2n)-th order coherence" with respect to the coherence phenomena associated with the correlation function (1.6).

*Acknowledgments*. Thanks are due to my fellow contributors — Prof. H.A. Ferwerda, J. Geist, Prof. B.J. Hoenders, Prof. H.G. Schmidt-Weinmar, Prof. A. Walther, and Prof. A. Zardecki — for finding time, energy, and enthusiasm to prepare their contributions, meeting the deadlines as well as they could in spite of numerous other commitments, and patiently considering the editor's suggestions. The active interest of Dr. H.K.V. Lotsch, who suggested the present volume, is gratefully acknowledged.

As a relative newcomer to the field of inverse problems in optics, the editor has profited from stimulating discussions and correspondence with many colleagues, most of whose names appear in the literature referenced throughout this volume. Particularly useful were the discussions during the Boston meeting of the Optical Society of America in Fall 1975 and the Fourth Rochester Conference on Coherence and Quantum Optics in 1977.

Thanks are due to Prof. L.M. Narducci for the trouble he took to establish the first contact between Prof. A. Walther and the editor.

Much of the editor's interest in the field was stimulated by suggestions of Dr. J.-F. Moser, head of the physics laboratory of LANDIS & GYR ZUG, who was kind enough to write the foreword of this book. The *Zentrale Forschung und Entwicklung* of LGZ has sponsored studies of theoretical problems in optical physics in the applied research program. Many colloquia and discussions were made possible by this department. Moreover, the continuous collaboration with Dr. B. Steinle, mathematician at LGZ, has been an important support for this work.

Allying competence with devotion to the task, Mrs. G. Baltes has provided valuable technical assistance in the editorial work.

## References


1.1 K. Chadan, P.C. Sabatier: *Inverse Problems in Quantum Scattering Theory* (Springer, New York, Heidelberg, Berlin 1977)
1.2 J.W. Goodman: "Synthetic-Aperture Optics", in *Progress in Optics*, ed. by E. Wolf, Vol.VIII (North-Holland, Amsterdam, London 1970) pp. 1-50
1.3 B.R. Frieden: "Evaluation, Design and Extrapolation for Optical Signals, Based on Use of the Prolate Functions", in *Progress in Optics*, ed. by E. Wolf, Vol. IX (North-Holland, Amsterdam, New York 1971) pp. 311-407
1.4 G.A. Viano: J. Math. Phys. *17*, 1160-1165 (1977)
1.5 F.T. Arecchi, V. Degiorgio: "Measurement of the Statistical Properties of Optical Fields", in *Laserhandbook*, ed. by F.T. Arecchi, E.O. Schulz-Dubois, Vol. 1 (North-Holland, Amsterdam, New York, Oxford 1972) pp. 191-264
1.6 C.W. Helstrom: "Quantum Detection Theory", in *Progress in Optics*, ed. by E. Wolf, Vol. X (North-Holland, Amsterdam, London 1972) pp. 289-369
1.7 G. Ross: Phil. Trans. Roy. Soc. (London) *268*, 177-200 (1970)
1.8 D. Gabor: "Light and Information", in *Proceedings of a Symposium on Astronomical Optics and Related Subjects*, ed. by Z. Kopal (North-Holland, Amsterdam 1956) pp. 17-30; and in *Progress in Optics*, ed. by E. Wolf, Vol. I (North-Holland, Amsterdam 1966) pp.109-153
1.9 G. Parry: "Speckle Patterns in Partially Coherent Light", in *Laser Speckle and Related Phenomena*, ed. by J.C. Dainty, Topics in Applied Physics, Vol. 9 (Springer, Berlin, Heidelberg, New York 1975) Sect.3.1
1.10 E. Jakeman, J.G. Whirter: J. Phys. A *9*, 785-797 (1976);
P.N. Pusey: J. Phys. D *9*, 1399-1409 (1976)
1.11 J.Y. Wang, R. Goulard: Appl. Opt. *14*, 862-871 (1975)
1.12 A. Ishimaru: Proc. IEEE *65*, 1030-1061 (1977) (review)
1.13 H.W. Carter, P.-C. Ho: Appl. Opt. *13*, 162-172 (1974)
1.14 H.P. Baltes, E. Hilf: *Spectra of Finite Systems* (Bibliographisches Institut, Zürich 1976)
1.15 A.-R.M. Zaghloul, R.M.A. Azzam, N.M. Bashara: "Inversion of the Nonlinear Equations of Reflection Ellipsometry on Film-substrate Systems", in *Ellipsometry*, ed. by N.M. Bashara, R.M.A. Azzam (North-Holland, Amsterdam, New York 1976) pp. 87-96
1.16 M. Bertolotti: "Photon Statistics", in *Photon Correlation and Light Beating Spectroscopy*, ed. by H.Z. Cummins, E.R. Pike (Plenum Press, New York, London 1974) pp. 41-74, Chap.5
1.17 A. Selloni, P. Schwendimann, A. Quattropani, H.P. Baltes: Open-system theory of photodetection: Dynamics of field and atomic moments. J. Phys. A*7*, 1427-38 (1978) Thermal effects in Photodetection. Scheduled for Phys. Rev. A*18* (1978)
1.18 E.T. Jaynes: Phys. Rev. *106*, 620-630 (1957)
1.19 R.S. Ingarden, A. Kossakowski: Bull. Acad. Polon. Sci., Sêr. Sci. Math. Astronom. Phys. *19*, 83-85 (1971)
1.20 P. Schwendimann, H.P. Baltes, A. Quattropani: Relevance of subpoissonian statistics to photon antibunching. Helv. Phys. Acta (to be published)
1.21 R. Kikuchi, B.H. Soffer: J. Opt. Soc. Am. *67*, 1656-1665 (1977)
1.22 L.M. Narducci, P.C. Colby, V. Bluemel, R.A. Tuft: Phys. Lett. *57A*, 204-206 (1976);
P.C. Colby, L.M. Narducci, V. Bluemel, J. Baer: Phys. Rev. A *12*, 1530-1538 (1975)
1.23 T. Hashimoto, Y. Murakami, H. Kawai: J. Polym. Sci. *13*, 1613-1631 (1975)
1.24 E. Schanda (ed.): *Remote Sensing for Environmental Sciences*, Ecological Studies, Vol. 18 (Springer, Berlin, Heidelberg, New York 1976)
1.25 E.D. Hinkley (ed.): *Laser Monitoring of the Atmosphere*, Topics in Applied Physics, Vol. 14 (Springer, Berlin, Heidelberg, New York 1976)
1.26 J.H. Shapiro: "Imaging and Optical Communication Through Atmospheric Turbulence", in *Laser Beam Propagation in the Atmosphere*, ed. by J.W. Strohbehn, Topics in Applied Physics, Vol. 25 (Springer, New York, Heidelberg, Berlin 1978)
1.27 R.J. Collier, C.B. Burckhardt, L.H. Lin: *Optical Holography* (Academic Press, New York, London 1971) Chap.19

1.28 G. Antes, D.L. Greenaway: Projektionsschirme mit optimalisierter Lichtausbeute und Fremdlichtdiskrimination. Helv. Phys. Acta (to be published)
1.29 R.J. Glauber: "Coherence and Quantum Detection", in *Quantum Optics*, ed. by R.J. Glauber (Academic Press, New York, London 1969) pp. 15-56, Chap.3
1.30 J.R. Klauder, E.C.G. Sudarshan: *Fundamentals of Quantum Optics* (Benjamin, New York, Amsterdam 1968) Chaps.1,8
1.31 L. Mandel, E. Wolf: Rev. Mod. Phys. *37*, 231-287 (1965) (review)
1.32 J. Peřina: *Coherence of Light* (Van Nostrand Reinhold, New York, Cincinnati, Toronto, Melbourne 1971)
1.33 J. Peřina, V. Peřinová, Z. Braunerová: Opt. Appl. *7/3*, 79-83 (1977)

# 2. The Phase Reconstruction Problem for Wave Amplitudes and Coherence Functions

H. A. Ferwerda

With 3 Figures

The phase retrieval problem arises in many different branches of physics. For instance, in x-ray structure determination, only the absolute magnitudes of the structure factors can be determined, the phase apparently gets lost. Scattering distributions (differential cross-sections) only yield the absolute square of the scattering amplitude, while the knowledge of its phase is indispensable for the determination of the structure of the scattering objects. A similar problem arises in structure determination with a light or an electron microscope (with particular emphasis on the latter application): the directly measurable quantity in this case is the intensity distribution in the image plane or some other plane in the microscope. This gives us the absolute square of the wave function[1] in this particular plane. In order to determine the object structure we also need the phase of the wave function. The greater part of this chapter will be devoted to the determination of the complex wave functions from one or more intensity distributions. In the final sections we shall briefly discuss another phase problem in optics, namely the phase problem of optical coherence theory: in experiments only the modulus of the mutual intensity function $W(\underline{r}_1,\underline{r}_2)$ is readily measurable, the phase of $W(\underline{r}_1,\underline{r}_2)$ has to be deduced indirectly, if possible.

It should be stressed that the solution of the phase retrieval problem, i.e., the calculation of the phase of a function from its modulus, is only possible when we know in advance that the complex function under consideration belongs to a particular class of functions. WALTHER and WOLF were the first researchers to raise this fundamental point [2.1,26]. In the case of optics we deal with band-limited functions.

In Section 2.1.1 we shall briefly explain the relevance of the phase problem for the object structure determination. A more elaborate discussion will be given in Chapter 3.

[1]Throughout this chapter we use the scalar theory of image formation both for the light and the electron microscope. The scalar complex amplitudes will be denoted as wave functions.

## 2.1 Phase Reconstruction for Wave Amplitudes

### 2.1.1 Relevance of the Phase Problem for Object Structure Determination

Let us consider an object which consists of a spatially varying distribution of the refractive index, $n(\underline{r})$. In order to determine this distribution we use quasi-monochromatic fully coherent radiation. For the sake of simplicity of presentation we have a plane wave incident on the distribution. If the phase retrieval problem has been solved, we know the wave function in a plane immediately behind the object, which has been denoted as "object plane". Let $u(\underline{r}_0)$ denote the wave function in the object plane and let $\exp(i\underline{k}\cdot\underline{r}_0)$ denote the undisturbed incident wave function in this plane ($\underline{r}_0$ denotes a vector in the object plane). If the z-axis is chosen along the direction of incidence, and if the object plane is given the z-coordinate zero, we find for the wave function in the object plane[2]

$$u(\underline{r}_0) = \text{const} \int_{-\infty}^{\infty} n(\underline{r}_0, z)\, dz \quad . \tag{2.1}$$

So the phase retrieval solution gives us a projection of the refractive index distribution on a plane perpendicular to the direction of incidence. By illuminating the object from different directions we obtain different projections of $n(\underline{r})$. If the number of projections is sufficiently large, we can construct the three-dimensional distribution $n(\underline{r})$ from these projections. These problems will be more generally and thoroughly discussed in Chapter 3.

We shall now return to the phase retrieval problem. In the next section we will derive the basic equations governing the phase problem.

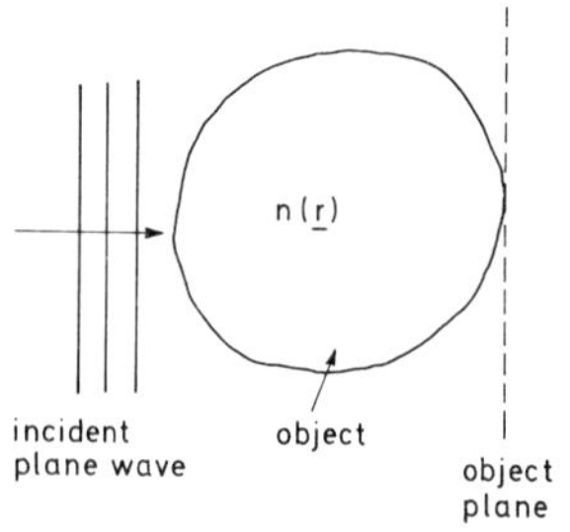


Fig. 2.1. Illustrating the notation

---

[2]In order to bring out explicitly the z-dependence of $n(\underline{r})$ we wrote this quantity as $n(\underline{r}_0,z)$. Here $\underline{r}_0$ is a vector perpendicular to the z-axis specifying the position of a point in the object plane.

## 2.1.2 Derivation of the Basic Equations Governing the Phase Problem

We first present a schematic arrangement of a general optical imaging process in Fig.2.2.

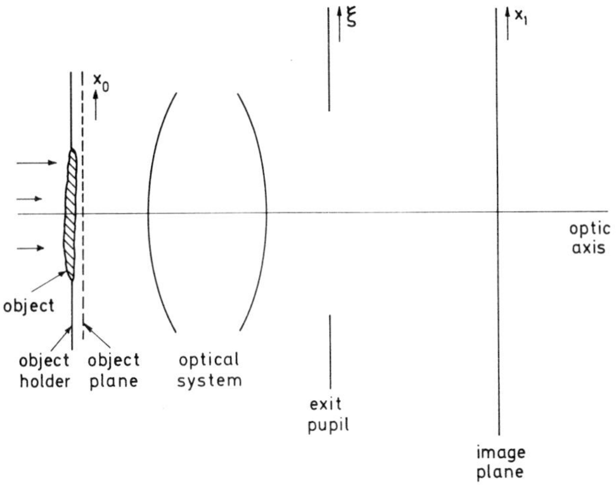


Fig. 2.2. Schematic arrangement of general imaging process

We assume quasi-monochromatic fully spatially coherent illumination. In light optics this assumption can be excellently approximated by using laser illumination. In the case of electron microscopy, the assumption is more problematic, as high coherence can be achieved only by using field emission guns in combination with diaphragms. The latter will reduce the available intensity.

The coordinates are defined as follows: the z-axis is chosen along the optic axis. By $\underline{r}_0 = (x_0, y_0)$ we denote the position of a point in the object plane, which is assigned the z-coordinate zero (the object plane is defined as the plane immediately behind the object perpendicular to the optic axis). The position of a point in the exit pupil is denoted by the vector $\underline{\rho} = (\xi, \eta)$, while a point in the image plane is described by $\underline{r}_1 = (x_1, y_1)$. Throughout this chapter we assume quasi-monochromatic, fully spatially coherent radiation with wavelength $\lambda$. For electron microscopy this means that we only consider image formation by the elastically scattered electrons (the inelastically scattered electrons undergo a change of wavelength upon scattering). Also full spatial coherence is only compatible with elastic scattering. The distance $\underline{r}_0$ will be measured in units $\lambda$, $\underline{\rho}$ in units F, where F is the focal length of the system and $\underline{r}_1$ is measured in units $M\lambda$, where M is the magnification. The following wave functions will be encountered: $u_0(\underline{r}_0)$ is the wave function in the object plane ("object wave function") and is the quantity which has to be ultimately reconstructed. The relation between $u_0(\underline{r}_0)$ and the object structure has been indicated in (2.1). Moreover $p(\underline{\rho})$ is the wave function in the exit pupil. Here $p(\underline{\rho})$ includes the aberrations of the optical system:

$$p(\underline{\rho}) = \int_{\sigma_0} d\underline{r}_0\, u_0(\underline{r}_0)\, \exp\{2\pi i[\phi(\underline{r}_0,\underline{\rho})+\underline{r}_0\cdot\underline{\rho}]\} \quad , \tag{2.2}$$

where $\phi(\underline{r}_0,\underline{\rho})$ is the aberration function in the exit pupil [2.2], and $\sigma_0$ is the transparent part of the object plane and is considered as the extent of the object. We calculated $p(\underline{\rho})$ assuming the traditional scalar Kirchhoff diffraction theory. If the imaging is isoplanatic [2.2], the aberration function $\phi(\underline{r}_0,\underline{\rho})$ does not depend on $\underline{r}_0$: $\phi(\underline{r}_0,\underline{\rho}) = \phi(\underline{\rho})$. In that case $\exp[2\pi i\phi(\underline{\rho})]$ can be taken outside the integral sign in (2.2), which now essentially expresses a Fourier transform relationship between $u_0(\underline{r}_0)$ and $p(\underline{\rho})$. This means that if $p(\underline{\rho})$ is the result of the phase retrieval solution, $u_0(\underline{r}_0)$ follows by Fourier inversion. For nonisoplanatic aberrations this inversion can also be done, as has been shown in [2.3], but the procedure becomes much more cumbersome. The relation between the wave function in the image plane, $u_i(\underline{r}_i)$ and the wave function in the exit pupil $p(\underline{\rho})$ is a Fourier transform relation

$$u_i(\underline{r}_i) = \int_{\sigma} d\underline{\rho}\; p(\underline{\rho})\exp(-2\pi i\underline{r}_1\cdot\underline{\rho}) \quad ; \tag{2.3}$$

where $\sigma$ is the aperture in the exit pupil [Ref.2.2, p.482]. We shall try to solve $p(\underline{\rho})$ from judiciously chosen intensity distributions. The intensity distribution in the image plane, $I_i(\underline{r}_i)$, is given by

$$I_i(\underline{r}_i) = |u_i(\underline{r}_i)|^2 \quad . \tag{2.4}$$

The intensity distribution in the exit pupil is given by $I_e(\underline{\rho}) = |p(\underline{\rho})|^2$. In order to avoid inessential complications (they are in any case inessential for isoplanatic imaging), we shall assume only one spatial dimension perpendicular to the optic axis.

### 2.1.3 General Considerations on the Phase Problem

In the introduction it has already been emphasized that the phase problem is only solvable if the complex functions under consideration belong to a particular class. That this is the case can be seen from (2.2) and (2.3): if the wave function $u_0(\underline{r}_0)$ has a finite support and is continuous, then by a well-known theorem [2.4] $p(\xi)$ is an entire function of $\xi$. In (2.3) only that part of $p(\xi)$ is used with $\xi \in \sigma$. The same theorem now states that $u_i(x_1)$ is an entire function of $x_1$. So the wave functions occurring in optics are entire functions. For this type of function the modulus and phase are related by a dispersion relation which will be derived by using the fact that $\log u_i(x_1)$ is an analytic function with branch points in the zeros of $u_i(x_1)$. We shall derive a dispersion relation for the image wave function $u(x)$ (from now on we drop the indices "i" and "1") along the lines set out by HOENDERS [2.5] to which the reader is referred for further details.

Let the extent of the exit pupil be given by $\alpha \leqq \xi \leqq \beta$. Equation (2.3) then reads

$$u(x) = \int_{\alpha}^{\beta} p(\xi)\exp(-2\pi i x\xi)d\xi \quad , \tag{2.5}$$

from which we easily deduce the asymptotic behavior of $u(x)$ as $x\to\infty$:

$$u(x) \sim \frac{-p(\alpha)e^{-2\pi i x\alpha}}{2\pi i x}\left[\frac{p(\beta)}{p(\alpha)}e^{-2\pi i x(\beta-\alpha)}-1\right] \quad . \tag{2.6}$$

From (2.6) it follows that the zeros of $u(x)$ are asymptotically distributed as

$$a_n \sim (\beta-\alpha)^{-1}\,(n+\gamma) \quad , \tag{2.7}$$

where $n$ is an integer and $\gamma$ is defined by

$$p(\alpha)^{-1}\,p(\beta) = \exp(2\pi i\gamma) \quad , \tag{2.8}$$

under the evident assumption $p(\alpha)p(\beta) \neq 0$. Let us discuss the case $\mathrm{Im}\,\gamma > 0$. Then, asymptotically, the zeros of $u(x)$ lie in the upper half plane (uhp) so that only a finite number of zeros is situated in the lower half plane (lhp). Let us denote these zeros by $a_1,a_2,\ldots,a_\ell$. It is also seen from (2.6) that these zeros are simple. Next we construct from $u(x)$ a function $w(x)$ which has no zeros in the lhp by "flipping" the zeros of $u(x)$ into their conjugate positions. This leads to the function

$$w(x) = u(x)\prod_{n=1}^{\ell}\frac{x-a_n^*}{x-a_n} \quad , \tag{2.9}$$

where $a_n^*$ denotes the complex conjugate of $a_n$. The product occurring in (2.9) is a so-called Blaschke-factor. As $w(x)$ now has no zeros in the lhp, $\log w(x)$ is regular there. We now consider the following contour integral

$$I(x,R) = (\pi i)^{-1}\int_C \frac{\ln[w(x')x'\exp(2\pi i\alpha x')]}{x'-x}dx' \quad , \tag{2.10}$$

where $C$ consists of the interval $[-R,R]$ along the real axis, indented into the lhp with semi-circles of vanishing radius around possible zeros of $w(x)$ on the real axis, and a semicircle of radius $R$ in the lhp with center at the origin. According to Cauchy's theorem $I(x,R) = 0$. On the other hand the contribution from the different parts of $C$ can be evaluated in the limit $R\to\infty$. This leads to the equation

$$\ln[w(x)x\,\exp(2\pi i\alpha x)] = \log[(2\pi i)^{-1}p(\alpha)]$$

$$-\frac{1}{\pi i}\,P\int_{-\infty}^{\infty}\frac{\ln[x'w(x')\exp(2\pi i\alpha x')]}{x'-x}dx' \quad , \tag{2.11}$$

where P denotes the Cauchy principal value. Equating the imaginary parts on both sides of (2.11) and using (2.9) we find the desired dispersion relation:

$$\phi(x) = \frac{1}{\pi} P \int_{-\infty}^{\infty} \frac{\ln|x'u(x')|}{x'-x} dx' + 2 \sum_{n=1}^{\ell} \arg(x-a_n) + \arg p(\alpha) - \frac{\pi}{2} - 2\pi\alpha x \quad , \tag{2.12}$$

where $\phi(x) = \arg u(x)$. The constants $\arg p(\alpha)$ and $\pi/2$ on the rhs of (2.12) have no physical consequence as the phase of $u(x)$ is determined up to an arbitrary constant. The linear term in x on the rhs of (2.12) reflects the fact that $u(x)$ is an entire function of order unity [Ref.2.4, Sect.7.4]. From the derivation leading to (2.12) we immediately deduce that the phase $\phi(x)$ can be uniquely deduced from $u(x)$ if $u(x)$ is known to possess only real zeros, a result which has already been obtained by WALTHER [2.1]. For the application of (2.12) we have to know the zeros of $u(x)$ in the lhp. This information is almost impossible to obtain from experiments. This can be understood as follows: the intensity $I(x) = u(x)\, u^*(x)$ for real values of x can be continued to complex values of x by

$$I(x) = u(x)\, u^*(x^*) \quad . \tag{2.13}$$

Now $I(x)$ is an analytic function in the whole complex x-plane. It is impossible to decide whether a complex zero of $I(x)$ is due to a zero of $u(x)$ or $u^*(x^*)$. Thus it is impossible to distinguish between $u(x)$ and $\tilde{u}(x) = u(x) \prod_{n=1}^{N} (x-a_n^*)(x-a_n)^{-1}$, as these functions have the same modulus on the real axis: $|\tilde{u}(x)| = |u(x)|$. The number of factors in the product occurring in the definition of $\tilde{u}(x)$ may be infinite, provided that the product converges. If N is the number of relevant complex zeros of $I(x)$, where the meaning of the adjective "relevant" will be explained in more detail in Section 2.1.6, we have an ambiguity in the solution of $u(x)$ from $I(x)$ of the order $2^N$ because we do not know whether a complex zero of $I(x)$ must be attributed to $u(x)$ or $u^*(x^*)$. From (2.7) we can derive that N is approximately equal to the number of degrees of freedom (Shannon number) of the image. Even the band limitation does not allow us to distinguish between $u(x)$ and $\tilde{u}(x)$, for both functions can easily be shown to have the same bandwidth [2.5,6]:

$$u(x) = \int_{\alpha}^{\beta} p(\xi) \exp(-2\pi i x\xi)\, d\xi \quad , \tag{2.14}$$

$$\tilde{u}(x) = \int_{\alpha}^{\beta} \tilde{p}(\xi) \exp(-2\pi i x\xi)\, d\xi \quad , \tag{2.15}$$

where $p(\xi)$ and $\tilde{p}(\xi)$ are related to each other by (see [2.5]):

$$\tilde{p}(\xi) = p(\xi) + 2\pi i \sum_{j=1}^{N} (2i)^j \sum_{\{j\}} \prod_{k\varepsilon\{j\}} |y_k|$$

$$\times \left\{ \sum_{\substack{m\varepsilon\{j\} \\ y_m>0}} \prod_{\substack{n \\ n\neq m}} \frac{a_m - a_n^*}{a_m - a_n} \int_{\alpha}^{\xi} p(\xi')\exp[-2\pi i a_m(\xi'-\xi)]d\xi' \right.$$

$$\left. + \sum_{\substack{m\varepsilon\{j\} \\ y_m<0}} \prod_{\substack{n \\ n\neq m}} \frac{a_m - a_n^*}{a_m - a_n} \int_{\xi}^{\beta} p(\xi')\exp[-2\pi i a_m(\xi'-\xi)]d\xi' \right\} \quad (\alpha<\xi<\beta) \quad . \tag{2.16}$$

Here $y_k = \mathrm{Im}\ a_k$ while $\{j\}$ denotes the set of all j-tuples contained in the set $(1,2,\dots,N)$. We shall explain that (2.16) offers the possibility to show that only one $p(\xi)$ can be correct. First of all we notice that $p(\xi)$ is an entire function because of (2.2). In the case of isoplanatic imaging $p(\xi)$ is essentially the Fourier transform of $u_0(\underline{x}_0)$, and so should vanish when $\xi \to \pm\infty$ along the real $\xi$-axis. Equation (2.16) holds in the whole complex $\xi$-plane because both sides of the equation are entire functions of $\xi$. Suppose $p(\xi)$ on the rhs of (2.16) is the correct solution. Then $\tilde{p}(\xi)$ cannot be a correct solution because $\tilde{p}(\xi)$ will diverge when $\xi \to \pm\infty$ along real values as the zeros $a_n$ have nonvanishing imaginary parts.

Phase retrieval from a single intensity distribution is thus possible if we know whether $\mathrm{Im}\ \gamma > 0$ or $< 0$, i.e., from (2.8) whether $|p(\alpha)^{-1}p(\beta)| < 1$ or $> 1$, respectively. It should be clear, however, that the procedure described above is more a proof of the existence of a unique solution of the phase problem than a procedure which is readily applicable to practical situations.

### 2.1.4 Greenaway's Proposal for Phase Recovery from a Single Intensity Distribution

Recently GREENAWAY [2.7] has formulated a proposal for phase recovery from a single intensity distribution. This aim can be achieved by judiciously placing an opaque stop across the exit pupil, so that $p(\xi)$ is only different from zero in the intervals $\alpha \leq \xi \leq \gamma$, $\delta \leq \xi \leq \beta$ with $\gamma < \delta$. While in the preceding section $p(\xi)$ had to vanish when $\xi \to \pm\infty$ along the real axis (to be established by analytical continuation!), we now have the more convenient condition that $p(\xi)$ has to vanish on $\gamma < \xi < \delta$, where the stop is located. Let us split $p(\xi)$ as

$$p(\xi) = q_1(\xi) + q_2(\xi) \quad , \tag{2.17}$$

where $q_1(\xi)$ vanishes outside $\alpha < \xi < \gamma$ and $q_2(\xi)$ vanishes outside $\delta < \xi < \beta$. Just as in the preceding section we consider two functions $u(x)$ and $\tilde{u}(x)$ which can be derived from each other by zero flipping. Equation (2.16) then yields for $\gamma < \xi < \delta$ [remember that $p(\xi)$ vanishes there]:

$$\tilde{p}(\xi) = 2\pi i \sum_{j=1}^{N} (2i)^j \sum_{\{j\}} \prod_{k\epsilon\{j\}} |y_k|$$

$$\times \left[ \sum_{\substack{m\epsilon\{j\} \\ y_m>0}} \prod_{\substack{n \\ n\neq m}} \frac{a_m - a_n^*}{a_m - a_n} \exp(2\pi i a_m \xi) Q_1(a_m) \right.$$

$$\left. + \sum_{\substack{m\epsilon\{j\} \\ y_m<0}} \prod_{\substack{n \\ n\neq m}} \frac{a_m - a_n^*}{a_m - a_n} \exp(2\pi i a_m \xi) Q_2(a_m) \right] \quad (\gamma<\xi<\delta) \quad , \tag{2.18}$$

where $Q_1(x)$ and $Q_2(x)$ are the Fourier transforms of $q_1(\xi)$ and $q_2(\xi)$. The question is whether apart from the correct solution $p(\xi)$ there are other acceptable solutions $\tilde{p}(\xi)$, which also vanish on $\gamma < \xi < \delta$. So we have to find out whether the rhs of (2.18) can be identically zero on $\gamma < \xi < \delta$. This is impossible if a finite number of zeros has been flipped, because the rhs of (2.18) is an entire function of $\xi$, which cannot vanish identically on $\gamma < \xi < \delta$ unless all $Q_1(a_m)$ and $Q_2(a_m)$ vanish. In the latter case we have, because of (2.5) and (2.17):

$$u(a_m) = Q_1(a_m) = Q_2(a_m) = 0 \quad . \tag{2.19}$$

This rather improbable situation might occur when $q_1(\xi)$ and $q_2(\xi)$ can be obtained from each other by translation over a distance $\sigma : q_1(\xi) = q_2(\xi+\sigma)$. This cannot occur when the stop $\gamma < \xi < \delta$ is placed asymmetrically in the aperture in the exit pupil. If $q_1$ and $q_2$ are not identical apart from a lateral shift, then an ambiguity $2^c$ remains, where c is the number of common zeros of $Q_1$ and $Q_2$. When all the zeros of $u(x)$ are flipped we obtain $u^*(x)$. Now $u(x)$ and $u^*(x)$ may have Fourier transforms which vanish in $\gamma < \xi < \delta$, unless the stop is placed asymmetrically in the aperture of the exit pupil.

Also the present proposal shows that phase retrieval can be achieved, in principle, from one intensity distribution, but may be very hard to apply in practical situations. The merit of the present proposal is that we do not have to know the location of the zeros of $u(x)$, apart from some special pathological situations. The proposal will not work for electron microscopy because of charging effects of the aperture stop which will introduce unwanted phase shifts.

### 2.1.5 The Method of Half-Plane Apertures for Semi-Weak Objects

Let us consider the case that the exit pupil is constructed such that only one half is transparent:

$$\begin{aligned} p(\xi) &\neq 0 \quad \text{for } 0 < \xi < \infty \\ p(\xi) &= 0 \quad \text{for } -\infty < \xi \leq 0 \quad . \end{aligned} \tag{2.20}$$

Then $u(x)$ is a causal transform [2.8] and its real and imaginary parts form a pair of Hilbert transforms:

$$\mathrm{Im}\, u(x) = \frac{-1}{\pi} P \int_{-\infty}^{\infty} \frac{\mathrm{Re}\, u(x')}{x'-x} dx' \qquad (2.21)$$

$$\mathrm{Re}\, u(x) = \frac{1}{\pi} P \int_{-\infty}^{\infty} \frac{\mathrm{Im}\, u(x')}{x'-x} dx' \quad . \qquad (2.22)$$

Let us first discuss the case of a *weak* object. An object will be called weak when the image wave function can be written as

$$u(x) = 1 + u_s(x) \quad , \qquad (2.23)$$

where $|u_s(x)| \ll 1$. Here $u_s(x)$ describes the scattered wave which is supposed to be so weak that in the formula for the intensity in the image plane,

$$I(x) = 1 + 2\,\mathrm{Re}\, u_s(x) + [\mathrm{Re}\, u_s(x)]^2 + [\mathrm{Im}\, u_s(x)]^2 \quad , \qquad (2.24)$$

the terms quadratic in $\mathrm{Re}\, u_s(x)$ and $\mathrm{Im}\, u_s(x)$ may be neglected with regard to $2\,\mathrm{Re}\, u_s(x)$:

$$I_{weak}(x) = 1 + 2\,\mathrm{Re}\, u_s(x) \quad . \qquad (2.25)$$

If this approximation is reliable, $\mathrm{Re}\, u_s(x)$ follows from $I_{weak}(x)$. Substitution of (2.23) in (2.21) yields

$$\mathrm{Im}\, u_s(x) = -\pi^{-1} P \int_{-\infty}^{\infty} \frac{\mathrm{Re}\, u_s(x')}{x'-x} dx' \quad . \qquad (2.26)$$

So now the complex image wave function has been determined from which $p(\xi)$ follows for $\xi > 0$. Taking a second exposure with the complementary diaphragm in the exit pupil such that $p(\xi) = 0$ for $\xi > 0$ gives $p(\xi)$ for $\xi < 0$. So now $p(\xi)$ is known on the full $\xi$-interval.

MISELL et al. [2.9] have proposed an improvement upon the weak object approximation by taking into account the quadratic term on the right hand side of (2.24). The procedure only works when the quadratic terms are not too large compared with the linear term. When this condition is fulfilled we shall call the object *semi-weak*. We shall solve $u_s(x)$ iteratively: in the first step of the iteration we neglect the quadratic terms in (2.24), which gives us $\mathrm{Re}\, u_s^{(1)}(x)$ (the superscript 1 denotes the step of the iteration). $\mathrm{Im}\, u_s^{(1)}(x)$ follows from

$$\mathrm{Im}\, u_s^{(1)}(x) = -\pi^{-1} P \int_{-\infty}^{\infty} \frac{\mathrm{Re}\, u_s^{(1)}(x')}{x'-x} dx' \quad . \qquad (2.27)$$

The iteration scheme continues as follows:

$$\mathrm{Re}\, u_s^{(n)}(x) = \frac{1}{2}\left\{I(x)-1-[\mathrm{Re}\, u_s^{(n-1)}(x)]^2-[\mathrm{Im}\, u_s^{(n-1)}(x)]^2\right\}$$

$$\mathrm{Im}\, u_s^{(n)}(x) = -\pi^{-1}\, P\int_{-\infty}^{\infty} \frac{\mathrm{Re}\, u_s^{(n)}(x')}{x'-x}dx' \quad . \tag{2.28}$$

The convergence of (2.28) has to be ascertained empirically. It turns out that no convergence is obtained when the quadratic terms are too strong. Taking two complementary half planes gives $p(\xi)$ for all $\xi$.

### 2.1.6 The Logarithmic Hilbert Transform: Methods for Circumventing Complications Due to Zeros

In Section 2.1.3 the logarithmic Hilbert transform (dispersion relation between the phase $\phi(x)$ and modulus $|u(x)|$) has been derived [(2.12)]. The difficulty encountered in practice is that the location of the zeros of $u(x)$ is unknown. In this section we shall discuss some cases where the detailed knowledge of these zeros is not necessary. They all refer to apertures in the exit pupil which lie on one side of the optic axis, e.g., $0 \leqq \alpha < \xi < \beta$. In that case $u(x)$ is a causal transform which means that Re $u(x)$ and Im $u(x)$ are related by a Hilbert transform [(2.12) and (2.22)]. When considering phase retrieval problems we are more interested in a relationship between the phase $\phi(x)$ of $u(x)$ and its modulus than in a relation between its real and imaginary part. This can be achieved by considering $\log u(x) = \log|u(x)| + i\phi(x)$. Unfortunately, when $u(x) \in L^2$ (square integrable functions) log $u(x)$ definitely is not because $\ln u(x) \to -\infty$ when $x \to \pm\infty$. BURGE et al. [2.10] have proposed methods for overcoming this difficulty and the difficulties associated with the zeros of $u(x)$ which become branch points of log $u(x)$. We shall briefly discuss a few methods. For a more detailed exposition the reader is referred to [2.10,11]. Let us consider the case that the aperture in the exit pupil lies above the optic axis: $0 \leqq \alpha < \xi < \beta$. Let $w(x)$ be the image wave function. From (2.6) we deduce that a coherent constant background C can be found such that the resulting image wave function

$$u(x) = C + w(x) \quad , \tag{2.29}$$

has no zeros in the lhp[3]. Without loss of generality, C can be chosen real and positive. From (2.6) we deduce that $\log\{[C+w(x)]C^{-1}\} \in L^2$ along every straight path in

[3]Strictly speaking a constant C is incompatible with the analyticity of $p(\xi)$ as C causes a $\delta$-function singularity in $p(\xi)$. It should cause no problem to remedy this lack of beauty.

the lhp parallel to the real axis. In that case $\phi(x)$ and $\log|u(x)|$ are related by a Hilbert transform because of Titchmarsh's theorem [Ref.2.8, p.34]

$$\phi(x) = \pi^{-1} P \int_{-\infty}^{\infty} \frac{\log|u(x')|}{x'-x} dx' \quad . \tag{2.30}$$

If $w(x)$ is the wave function to be reconstructed, we can determine the phase of $w(x)$ from

$$\tan \arg w(x) = \frac{|u(x)| \sin\phi(x)}{|u(x)| \cos\phi(x) - C} \quad . \tag{2.31}$$

$$|w(x)| = [C^2 + |u(x)|^2 - 2C\, u(x) \cos\phi(x)]^{1/2} \quad . \tag{2.32}$$

Another method closely resembling the preceding one is based on Rouchê's theorem [Ref.2.4, pp.119,120] and has also been proposed by BURGE et al. [2.10]: *Phase retrieval with a reference function*. We now add a spatially varying reference beam $R(x)$ coherently to the image wave function $u(x)$. $R(x)$ has to satisfy the following conditions:

a) $R(x)$ is regular in the lhp

b) $R(x)$ has no zeros in the lhp

c) $|u(x)| < |R(x)|$ on the real axis and any semicircle $|x| = \rho$ $(\rho \to \infty)$ in the lhp.

According to Rouchê's theorem $u(x) + R(x)$ then has no zeros in the lhp. So $\log[u(x)+R(x)]$ is regular there. We shall now derive a logarithmic Hilbert transform relation for $u(x) + R(x) = F(x)$. To this end we consider the contour integral

$$\int_C \frac{\log F(x') - \sum\limits_{j=0}^{n-1} \log^{(j)} F(0)\, x'^j (j!)^{-1}}{x'^{n+1}(x'-x)} dx' \quad . \tag{2.33}$$

Here $\log^{(j)} F(0) = [(d/dx)^j \log F(x)]_{x=0}$, and $n$ is a positive integer, $C$ is the closed contour consisting of the segment $-\rho \leq x' \leq \rho$ along the real axis, indented into the lhp at $x' = x$ and $x' = 0$ and the semicircle $|x'| = \rho$ in the lhp. As $F(x)$ is an entire function of order unity [as was already remarked in connection with (2.12)], $\log F(x)$ is bounded at infinity by a polynomial. When writing down (2.33) the bounding polynomial was assumed to have the degree $n$. In the case of optics $n$ may be chosen unity. The contribution of the semicircle of $C$ vanishes in the limit $\rho \to \infty$. Application of the calculus of residues gives after a straightforward calculation:

$$\arg F(x) = \sum_{j=0}^{n} x^j \operatorname{Im} K_j - \pi^{-1} x^{n+1} P \int_{-\infty}^{\infty} \frac{\log|F(x')| - \sum\limits_{j=0}^{n-1} x'^j \operatorname{Re} K_j}{x'^{n+1}(x'-x)} dx' \quad , \tag{2.34}$$

where we abbreviate

$$K_j \equiv (j!)^{-1} \log^{(j)} F(0) \quad . \tag{2.35}$$

In principle, Re $K_j = (j!)^{-1} \log^{(j)} |F(0)|$ can be determined from experiment, though the determination of derivatives is very inaccurate. In contrast, Im $K_j = \arg^{(j)} F(0)$ is unknown and represents an unknown parameter [subtraction constant in the dispersion relation (2.34)]. Obviously, the number of subtraction constants should be kept as low as possible.

BURGE et al. [2.10] have also made attempts to find the location of the complex zeros on the image intensity I(x). The number of zeros to be considered depends on the extent of the interval where u(x) has to be known and on the noise in the measurement. This statement is based upon a theorem due to TITCHMARSH [2.12] which proves that the modulus of an entire function on the real axis is determined by its zeros $a_i = \alpha_j \exp(i\theta_j)$ according to

$$|u(x)| = |u(0)| \prod_{j=1}^{\infty} |1 - x\alpha_j^{-1} \exp(-i\theta_j)| \quad . \tag{2.36}$$

Thus zeros far from the interval where u(x) has to be determined are uninteresting: their effect gets lost in the noise. A procedure for determining the positions of the relevant zeros has been described in [2.10].

The methods discussed so far make extensive use of analyticity properties. Their merit lies mainly in the fact that they tell whether the phase problem admits a solution. Their applicability to actual practical situations still has to be shown. In the following sections we will use more input information than is strictly necessary according to the results derived in the preceding sections: we shall make use of at least two intensity distributions, which can be two defocused images or the intensity distributions in image plane and exit pupil. As the phase has to be obtained numerically, we also have to be aware of the possible existence of nonanalytic solutions, which may inadvertently be produced by the computer, and which are unacceptable from a physical point of view. In the following sections we shall review successively phase retrieval for strong objects from two defocused images, phase retrieval for strong objects from intensity distributions in image plane and exit pupil, and phase retrieval for semi-weak objects from two defocused images with or without the intensity distribution in the exit pupil.

### 2.1.7 Phase Retrieval for Strong Objects from Two Defocused Images

In this section we shall treat the phase retrieval from two intensity distributions of two defocused images. An iterative solution, similar to the iterative scheme of the Gerchberg-Saxton algorithm to be discussed in Section 2.1.8, has been given by MISELL [2.13]. Even if the iteration scheme converges, we do not know whether the produced solution is acceptable, as long as the uniqueness of the solution has not been established. For that reason FERWERDA and collaborators [2.14-16] have studied the uniqueness of the present phase retrieval problem and have formulated algorithms which are more transparent mathematically than the one proposed by MISELL [2.13].

We shall first derive the basic formulas. Again we shall restrict ourselves to one transverse spatial dimension. $I(x_1)$ denotes the intensity in the image plane, $p(\xi)$ is the wave function in the exit pupil. In order to refer all quantities to one plane, in this case the plane of the exit pupil, we take the shifted Fourier transform of $I(x_1)$ (the shift is introduced for mathematical convenience):

$$f(\xi) = \int_{-\infty}^{\infty} I(x_1)\exp[2\pi i(\beta+\xi)x_1]dx_1 \quad . \tag{2.37}$$

Using (2.3) and (2.4) we find [2.14]

$$f(\xi) = \int_{\xi}^{\beta} p(\xi')p^*(\xi'-\xi-\beta)d\xi' \quad . \tag{2.38}$$

Here $\beta$ measures the width of the exit pupil: $-\beta < \xi < \beta$. We have two intensity distributions at our disposal, corresponding to two defocused images: $I_j(x_1)$, $j = 1,2$. If the corresponding defocusings are $\Delta z_j$ $(j=1,2)$ we find two equations [Ref.2.2, p.462, 2.14]

$$f_j(\xi) = \int_{\xi}^{\beta} p(\xi')p^*(\xi'-\xi-\beta) \exp\left\{2\pi i\Delta_j[\xi'^2-(\xi'-\xi-\beta)^2]\right\}d\xi' \quad , \quad j = 1,2 \quad , \tag{2.39}$$

where

$$\Delta_j = \pi\lambda^{-1}\Delta z_j \quad . \tag{2.40}$$

So $p(\xi)$ in (2.39) is the wave function in the exit pupil corresponding to the wave function in the Gaussian image plane, and $\lambda$ is the wavelength of the radiation.

It can easily be checked that one intensity distribution does not admit a unique solution: if $p(\xi)$ is a solution of (2.38) then also $p^*(-\xi)$. From the discussion in Section 2.1.3 it follows that there are only two solutions, one corresponding to $|p(\beta)[p(-\beta)]^{-1}| < 1$, and the other to $|p(\beta)[p(-\beta)]^{-1}| > 1$. When calculating $p(\xi)$ along these lines we have to guarantee analyticity during every step of the computation, which is impossible on the computer. In [2.14] the uniqueness of the solution has been established by DRENTH et al. directly exploiting the analyticity of $p(\xi)$. In a subsequent paper HUISER and FERWERDA [2.15] have established the uniqueness of the solution of (2.39) when the solution is differentiable almost everywhere. This gives more confidence in the solution which is ultimately produced by the computer. In the proof referred to above $p(\xi)$ and $p^*(\xi)$ have been treated as independent, i.e., the complex conjugation relationship has not been exploited. Sufficient conditions for proving the uniqueness are:

1) $p(\beta)$ is known and different from zero and $p^*(-\beta) \neq 0$.

2) $\sin\Delta(\xi^2-\beta^2) \neq 0$ which poses a restriction on the allowable defocusing. The condition can be met for every $\xi \in (-\beta,\beta)$ if

$$\Delta = \Delta_1 - \Delta_2 < \pi\beta^{-2} \quad . \tag{2.41}$$

It might seem highly impractical to demand the value of the wave function in one of the end points of the exit pupil. In principle the determination is possible, because $p(\beta)$ can be chosen to be real and positive. Notice from (2.39) that $p(\xi)$ can only be determined up to a constant phase factor. This freedom can be used to choose the overall phase such that $p(\beta) > 0$. Then $p(\beta)$ is found as the square root of the intensity in $\xi = \beta$. We shall show below that this practical difficulty can be elegantly solved.

We shall now present a direct method for solving $p(\xi)$ from (2.39). Another, more elaborate method, the Newton-Kantorovich algorithm, has been discussed in [2.16]. This algorithm is much more time-consuming, however.

The direct method proceeds by writing the integral equations into a set of algebraic equations. This is done by approximating the integrals by Riemann sums. To this end we divide the interval $-\beta < \xi < \beta$ by the sampling points:

$$\xi = \beta,\ \beta - h,\ \ldots\ldots.0,\ -\beta + h,\ -\beta \quad . \tag{2.42}$$

For convenience we suppose that $\xi = 0$ occurs as a sampling point. The length of the sampling interval is chosen in accordance with the Whittaker-Shannon theorem [2.17]: $h = 2\beta/\sigma_0$, where $\sigma_0$ is the width of the object. Since $p^*(-\beta)$ is assumed to be known[4], we can begin by writing down an equation for $p(\beta-h)$ which follows from the Riemann sum for $f_j(\beta-h)$ in (2.29):

$$f_j(\beta-h) = p(\beta-h)\ p^*(-\beta)\ \exp\left\{2\pi i\Delta_j[(\beta-h)^2-\beta^2]\right\} \quad . \tag{2.43}$$

Now $p(\beta-h)$ follows from (2.43). For $\xi = \beta - 2h$ we find:

$$f_j(\beta-2h) = p(\beta-2h)\ p^*(-\beta)\ \exp\left\{2\pi i\Delta_j[(\beta-2h)^2-\beta^2]\right\} + p(\beta-h)\ p^*(-\beta+h) \quad . \tag{2.45}$$

---

[4]This procedure is to be preferred to the one used in [2.16] where the trapezoidal rule for approximating an integral has been used. This requires knowledge of $p(\beta)$ *and* $p^*(-\beta)$. Here $p^*(-\beta)$ follows from $p(\beta)$ by the relation

$$f_j'(\beta) = -p(\beta)\ p^*(-\beta) \tag{2.44}$$

which causes problems in practice.

Equation (2.45) constitutes two linear equations in the unknown $p(\beta-2h)$ and $p^*(-\beta+h)$. The determinant of the coefficients does not vanish when (2.41) is satisfied. In this way we obtain a set of values:

$$\begin{aligned} &p(\beta-h),\ p(\beta-2h),\ldots,\ p(0),\ldots,\ p(-\beta+h) \\ &\qquad p^*(-\beta+h),\ldots,\ p^*(0),\ldots,\ p^*(\beta-h) \quad . \end{aligned} \tag{2.46}$$

Notice that during the calculation we never used the complex conjugation relationship between $p(\xi)$ and $p^*(\xi)$. When $p^*(-\beta)$ has the correct value, the complex conjugation relationship between $p(\beta-nh)$ and $p^*(\beta-nh)$ $(n=1,2,\ldots,N)$ is automatically satisfied. Unfortunately it is virtually impossible to obtain an accurate value for $p^*(-\beta)$. Therefore we start the calculation with an arbitrary nonzero value for $p^*(-\beta)$. This will undoubtedly destroy the complex conjugation-relationship between $p$ and $p^*$. In the uniqueness proof given in [2.14], $p$ and $p^*$ have been treated as independent. So in fact it has been shown there that the equations

$$f_j(\xi) = \int_{\xi}^{\beta} p(\xi')\, q(\xi'-\xi-\beta)\, \exp\left\{2\pi i \Delta_j [\xi'^2-(\xi'-\xi-\beta)^2]\right\} d\xi' \quad , \quad j = 1,2 \tag{2.47}$$

have a unique solution $\{p(\xi),q(\xi)\}$ under the conditions stated before (2.41). When we apply this result to our problem we see that a wrong value of $p^*(-\beta)$ will lead to a solution $\{p(\xi),q(\xi)\}$ which differs by a *constant* factor from the correct solution, for which $q = p^*$. If the solution based on a guessed value of $p^*(-\beta)$, say $\tilde{p}^*(-\beta)$, is denoted by $\{\tilde{p}(\xi),\tilde{q}(\xi)\}$ we have

$$\tilde{p}(\xi) = \chi p(\xi) \quad ; \quad \tilde{q}(\xi) = \chi^{-1} p^*(\xi) \quad , \tag{2.48}$$

where the second part of the equation follows from (2.47). Here $\chi$ is an unknown constant, equal to $\tilde{p}^*(-\beta)/p^*(-\beta)$ which can be found from (2.39) for $\xi = -\beta$:

$$f_j(-\beta) = \int_{-\beta}^{\beta} p(\xi')\, p^*(\xi')\, d\xi' = |\chi|^{-2} \int_{-\beta}^{\beta} \tilde{p}(\xi')[\tilde{p}(\xi')]^* d\xi' \quad , \tag{2.49}$$

where at this stage we used the complex conjugation relationship between $p$ and $p^*$. As $\tilde{p}(\xi)$ is known $|\chi|$ can be solved from (2.49). The argument of $\chi$ remains undetermined; this reflects the constant phase ambiguity. This procedure is much more accurate than trying to obtain $p^*(-\beta)$ by a direct measurement.

The algorithm described above has one drawback: the results are very sensitive to noise in the input data as the errors propagate through the different steps of the calculation. Another method which does not suffer from this disadvantage, but which applies only to semi-weak objects, will be discussed in Section 2.1.9.

### 2.1.8 Phase Retrieval from the Intensity Distributions in Exit Pupil and Image Plane

This method was first proposed by GERCHBERG and SAXTON [2.18,19]. The attempt is made to compute the complex wave function in the exit pupil from the intensities in image plane and exit pupil. The uniqueness of the solution has been investigated in a series of papers [2.20-22]. The mathematical problem to be solved is the construction of a complex function given its modulus and the modulus of its Fourier transform. In [2.20] the uniqueness problem has been studied for the case that $p(\xi)$ is an integral function. The solution has been shown to be unique unless the wave function is symmetric. In the latter case the solutions are $p(\xi)$ and $p^*(\xi)$ which shows that the ambiguity is rather harmless in this case. As the analyticity of the solution cannot be implemented in a numerical procedure, we have also to worry about the existence of other, nonanalytic, solutions. In [2.21] the existence of such solutions has been investigated. A criterion has been formulated which is sufficient for establishing the uniqueness of the solution. Unfortunately, this criterion depends on the function which still has to be solved, and explains at best a posteriori why no unique solution could be obtained in a particular case. Numerical study examples have confirmed the theory [2.23]. SCHISKE [2.22] has constructed explicit examples which show the nonuniqueness of the solution. For these reasons this method of phase retrieval is of doubtful practical usefulness.

With these limitations in mind we shall review some algorithms which have been proposed. In [2.18] GERCHBERG and SAXTON derive quadratic equations which they solve iteratively. Though numerical practice seems to indicate that always the same solution is obtained, the authors confess that the uniqueness of the solution is unknown. In a subsequent paper [2.19] the same authors present another algorithm which has become known as *the* Gerchberg-Saxton algorithm. The algorithm is started by making an Ansatz for the complex wave function in the exit pupil: the modulus is known from the intensity distribution in the exit pupil, the phase is produced by a random number generator from a uniform density distribution between $-\pi$ and $+\pi$. With this trial wave function the image wave function is computed with a fast Fourier transform (FFT). The modulus of the computed image wave function is replaced by the value following from the intensity distribution in the image plane while the computed phase is retained. Next an inverse FFT is used to compute $p(\xi)$ of which only the phase is retained while the modulus is changed into its correct value, etc. So the iterative procedure is

$$u^{(n)}(x) = \int_{-\beta}^{\beta} p^{(n-1)}(\xi)\exp(-2\pi i x\xi)d\xi \quad , \tag{2.50}$$

$$p^{(n)}(\xi) = \int_{-\infty}^{\infty} u^{(n)}(x)\exp(2\pi i x\xi)dx \quad . \tag{2.51}$$

In many cases this iterative procedure converges. The significance of the produced solution remains obscure as long as uniqueness is not guaranteed. From numerical experiments we have learned [2.23] that the solution depends on the choice for the trial function for the wave function in the exit pupil. As it is very difficult to understand what goes on mathematically, several researchers produced more transparent algorithms. A direct method has been presented in [2.23]. Another version of the direct method, which is an adaptation of a method developed by DALLAS [2.24], will be discussed below.

The fact that u(x) is a band-limited function according to (2.3) allows us to write down a Whittaker-Shannon expansion for u(x) [2.17]:

$$u(x) = \sum_{n=-\infty}^{\infty} u[(2\beta)^{-1}n]\mathrm{sinc}(2\beta x-n) \quad , \tag{2.52}$$

where $\mathrm{sinc}\, x = (\pi x)^{-1} \sin(\pi x)$. The wave function $p(\xi)$ can be calculated according to:

$$p(\xi) = \sum_{m=-\infty}^{\infty} (2\beta)^{-1} u[(2\beta)^{-1}m]\exp[-2\pi i m\xi(2\beta)^{-1}] \quad . \tag{2.53}$$

The quantities to be calculated are $u[(2\beta)^{-1}n]$, where the input consists of the intensities in exit pupil and image plane. In order to reduce all quantities to quantities in the image plane, we take the Fourier transform of the intensity distribution in the exit pupil:

$$C(y) = \int_{-\beta}^{\beta} |p(\xi)|^2 \exp(2\pi i y\xi)d\xi \quad . \tag{2.54}$$

Substituting $p(\xi) = \int_{-\infty}^{\infty} u(x)\exp(2\pi i x\xi)dx$ yields

$$C(y) = \int_{-\infty}^{\infty} dx\, u(x)u^*(x-y) \quad , \tag{2.55}$$

where we have carried out the calculation as if $\beta = \infty$ which is a good approximation if the diaphragm in the exit pupil does not intercept much energy. Substituting (2.52) into (2.55), and using the formula

$$\int_{-\infty}^{\infty} \frac{\sin m(x-\xi)}{x-\xi} \frac{\sin n(x-\eta)}{x-\eta} dx = \pi \frac{\sin m(\xi-\eta)}{\xi-\eta}$$

valid if $m \geqq n > 0$ (see [Ref.2.4, p.153, problem 12]) we arrive at

$$C(y) = \sum_{n,k=-\infty}^{\infty} (2\beta)^{-1} u_n u^*_{n-k}\, \mathrm{sinc}(2\beta y-k) \quad , \tag{2.56}$$

where we abbreviated $u[(2\beta)^{-1}n] = u_n$. As C(y) is band-limited with bandwidth $2\beta$ as follows from (2.54), C(y) can be expanded according to

$$C(y) = \sum_{k=-\infty}^{\infty} C_k \operatorname{sinc}(2\beta y-k) \quad , \tag{2.57}$$

where $C_k = C[(2\beta)^{-1}k]$. Comparing (2.56) and (2.57) we obtain the following set of algebraic equations:

$$2C_k = \sum_{n=-\infty}^{\infty} u_n u^*_{n-k} \quad . \tag{2.58}$$

As the image has essentially a finite extent, only a finite number of sampling points has to be considered:

$$-N \leqq n \leqq N \quad , \tag{2.59}$$

where N is the Shannon number (number of degrees of freedom) of the image. Furthermore, because of the symmetry property $C(y) = C^*(-y)$ which follows immediately from (2.54), we can restrict ourselves to values of $C_k$ with $k \geqq 0$: $k = 0,1,2,\ldots,2N$. The unknown coefficients $u_n$ can be solved successively from (2.58), which because of the finite extent of the image reads:

$$2\beta C_k = \sum_{n=-N+k}^{N} u_n u^*_{n-k} \quad . \tag{2.60}$$

For $k = 2N$, (2.60) becomes

$$2\beta C_{2N} = u_N u^*_{-N} \quad . \tag{2.61}$$

The constant-phase ambiguity allows us to choose $u_N$ real and positive. As $|u_N|$ follows from experiment $u_N$ is completely known, $u^*_{-N}$ and so also $u_{-N}$ is solved from (2.61). Taking $k = 2N - 1$ in (2.60) gives

$$2\beta C_{2N-1} = u_{N-1} u^*_{-N} + u_N u^*_{-N+1} \quad , \tag{2.62}$$

from which $u_{N-1}$ and $u^*_{-N+1}$ have to be solved (their moduli are known). From a simple geometric picture we immediately see that (2.62) in general has two solutions. This procedure is continued until all unknowns have been determined, which is the case when $k = N$ has been reached. The number of solutions may be as high as $2^N$. The remaining equations (2.60) for $k = 0,\ldots,N$ can be used to reduce the number of solutions.

### 2.1.9 Phase Retrieval from Two Defocused Images for Semi-Weak Objects

It was noticed in Section 2.1.7 that the direct method of phase retrieval for strong objects is very sensitive to noise. In particular in electron microscopy this presents a great practical difficulty because we cannot build up sufficient statistics due to radiation damage of the specimen by the electron beam. As the specimens of electron microscopy are as a rule either weak or semi-weak, we can exploit this circumstance for developing an algorithm which is less sensitive to noise.

Let us consider a weakly scattering object which is illuminated by a plane wave. The wave function in the exit pupil can then be written as

$$p(\xi) = c\delta(\xi-a) + p_1(\xi) \quad , \tag{2.63}$$

where a describes the direction of the incident beam and c describes the strength of that direction in the exit pupil [in practice $\delta(\xi-a)$ will become a sinc-function because of the finite extent of the object]. When writing down (2.63) we neglected spherical aberration. This restriction is not expected to affect the theory in a fundamental way. For the image wave function $u(x)$ we find from (2.63) using (2.3):

$$u(x) = c\ \exp(-2\pi iax) + \int_{-\beta}^{\beta} d\xi p_1(\xi)\ \exp(-2\pi ix\xi) \quad , \tag{2.64}$$

from which we calculate the intensity distribution $I(x) = |u(x)|^2$. For our calculations we shall need the Fourier transform of $I(x)$:

$$\hat{i}(\xi) = \int_{-\infty}^{\infty} I(x)\exp(2\pi ix\xi)dx \quad . \tag{2.65}$$

Two defocused images give the following equations:

$$\begin{aligned}\hat{i}_j(\xi) = {} & |c|^2\delta(\xi) + c^*p_1(\xi+a)\exp\left\{2\pi i\Delta_j[a^2-(\xi+a)^2]\right\} \\ & + c\ p_1^*(a-\xi)\exp\left\{2\pi i\Delta_j[(a-\xi)^2-a^2]\right\} \\ & + \int_{-\beta+\xi}^{\beta} p_1^*(\xi')p_1(\xi'-\xi)\exp\left\{2\pi i\Delta_j[\xi'^2-(\xi'-\xi)^2]\right\}d\xi' \quad , \\ & \text{for } 0<\xi<2\beta,\ j = 1,2 \\ & \quad -\beta<a<\beta \quad ,\end{aligned} \tag{2.66}$$

from which c and $p_1(\xi)$ have to be determined. If we assume that the object is semi-weak in the exit pupil, by which we mean that the terms linear in $p_1$ [2nd and 3rd term on the rhs of (2.66)] dominate over the quadratic term in (2.66), we can take this term into account in a convergent iterative scheme. Equation (2.66) is discretized by applying the Whittaker-Shannon sampling theorem to $p_1(\xi)$. To this end we define

$$\bar{\hat{i}}_j(\xi) = \hat{i}_j(\xi) - |c|^2\delta(\xi) \quad , \tag{2.67}$$

where we have to keep in mind that $\delta(\xi)$ is a sinc-like function. Let us assume that $a = \ell h$ is a sampling point ($h$ is the sampling distance $\sigma_0^{-1}$). Further $p_1(\xi)$ is defined such that

$$p_1(\ell h) = 0 \quad . \tag{2.68}$$

For the sake of simplicity of notation we abbreviate $p_1(\ell h) = p_{1,\ell}$. Equation (2.66) now reads when restricted to the sampling points:

$$\begin{aligned}\bar{\hat{i}}_{j,n} &= c\, p_{1,n+\ell} \exp\left\{2\pi i \Delta_j h^2[\ell^2-(n+\ell)^2]\right\} \\ &\quad + c\, p^*_{1,\ell-n} \exp\left\{2\pi i \Delta_j h^2[(\ell-n)^2-\ell^2]\right\} + \sum_m p^*_{1,m}\, p_{1,m-n}\, \gamma_{mn,j} \quad ,\end{aligned} \tag{2.69}$$

where

$$\begin{aligned}\gamma_{mn,j} &= \int_{-\beta+nh}^{\beta} \mathrm{sinc}[h^{-1}(\xi'-mh)]\, \mathrm{sinc}[h^{-1}(\xi'-nh-mh)] \\ &\quad \times \exp\left\{2\pi i \Delta_j[\xi'^2-(\xi'-nh)^2]\right\} d\xi' \quad .\end{aligned} \tag{2.70}$$

The overall phase has been chosen such that $c > 0$. The equations (2.69) are solved iteratively (a superscript denotes the step of the iteration):

$$\begin{aligned}\bar{\hat{i}}_{j,n} &= c^{(\nu)}\, p^{(\nu)}_{1,n+\ell} \exp\left\{2\pi i \Delta_j h^2[\ell^2-(n+\ell)^2]\right\} \\ &\quad + c^{(\nu)}\, p^{*(\nu)}_{1,\ell-n} \exp\left\{2\pi i \Delta_j h^2[(\ell-n)^2-\ell^2]\right\} + \sum_m p^{*(\nu-1)}_{1,m}\, p^{(\nu-1)}_{1,m-n}\, \gamma_{mn,j} \quad ,\end{aligned} \tag{2.71}$$

where the iteration is started by taking $p^{(0)}_{1,m} = p^{*(0)}_{1,m} = 0$, and $c^{(1)}$ is determined from

$$\hat{i}_{j,0} - c^{(1)^2}\delta(0) = 0 \quad . \tag{2.72}$$

The iteration for $c$ is continued by substituting $n = 0$ in (2.71) and using (2.68):

$$\hat{i}_{j,0} - c^{(\nu)2}\,\delta(0) = \sum_m \left|p^{*(\nu-1)}_{1,m}\right|^2 \gamma_{m0,j} \quad . \tag{2.73}$$

From (2.71) we can solve $p^{(\nu)}_{1,n+\ell}$ and $p^{*(\nu)}_{1,\ell-n}$ when $p_1^{(\nu-1)}$ and $p_1^{*(\nu-1)}$ etc. are known, provided that the determinant of the equations differs from zero, which is guaranteed over the whole interval when (2.41) is satisfied. The algorithm discussed above strongly recalls the off-axis holography technique [Ref.2.17, Sect.8.4]. This can be most clearly seen from (2.66) when the nonlinear term on the rhs is ignored. If the direction of illumination is such that $a = \pm\beta$, $p_1(\xi)$ can be calculated from one exposure.

In the case of electron microscopy where it is extremely important to extract as much information as possible from the scattered electrons, another method has recently been proposed by VAN TOORN et al. [2.25], where apart from two or more defocused images also the intensity distribution in the exit pupil is used. This procedure has been shown to give a substantial reduction in the dose necessary for reaching a sufficiently accurate solution.

## 2.2 Phase Reconstruction for Coherence Functions

### 2.2.1 Phase Determination of Optical Coherence Functions

In the preceding sections we discussed the phase retrieval for complex amplitudes. In this section we shall discuss the related problem of the phase determination of the mutual coherence function $\Gamma(\underline{r}_1,t_1;\underline{r}_2,t_2)$ defined by the ensemble average [Ref. 2.2, p.500]:

$$\Gamma(\underline{r}_1,t_1;\underline{r}_2,t_2) = \langle u(\underline{r}_1,t_1)u^*(\underline{r}_2,t_2)\rangle \quad , \tag{2.74}$$

where $u(\underline{r},t)$ denotes the stochastic complex amplitude at the point $\underline{r}$, at time t. If the field is temporally stationary $\Gamma$ only depends on the time difference $t_1 - t_2 = t$ and (2.74) becomes:

$$\Gamma(\underline{r}_1,t_1;\underline{r}_2,t_2) = \Gamma(\underline{r}_1,\underline{r}_2;t) \quad . \tag{2.75}$$

Equation (2.75) certainly holds for quasimonochromatic radiation.

We can distinguish two types of phase problems for optical coherence functions:

I) *The phase problem of temporal coherence*: $\underline{r}_1$ and $\underline{r}_2$ are fixed while t is variable ($\underline{r}_1$ and $\underline{r}_2$ may even coincide). This situation occurs in the Michelson interferometer [Ref.2.2, p.506,507] when the spectral density $g(\omega)$ of a spectral line has to be determined. Here $g(\omega)$ has to be solved from:

$$\gamma_{11}(t) = \int_0^\infty g(\omega)\exp(-i\omega t)d\omega \quad , \tag{2.76}$$

where

$$\gamma_{11}(t) = [I(\underline{r}_1)]^{-1}\,\Gamma(\underline{r}_1,\underline{r}_1;t) \quad , \tag{2.77}$$

and where $I(\underline{r}_1)$ is the intensity in $\underline{r}_1$. The spectral density $g(\omega)$ can only be obtained if the complex quantity $\gamma_{11}(t)$ is known. In principle this is possible because $|\gamma_{11}(t)|$ can be obtained from the visibility of the interference fringes in a Michelson interferometer [Ref.2.2, p.505] and the real part of $\gamma_{11}(t)$ follows from the interference pattern [Ref.2.2, Sect.10.3.1]. The detailed measurement of the interference pattern is in practice very difficult to carry out, so that $|\gamma_{11}(t)|$ remains the only experimentally accessible quantity. Thus we have to calculate $\gamma_{11}(t)$ from $|\gamma_{11}(t)|$. This problem has been discussed by WOLF [2.26], DIALETIS [2.27], and NUSSENZVEIG [2.28]. The same techniques are employed which have already been reviewed in Section 2.1.3.

II) *The phase problem of spatial coherence*: $\underline{r}_1$ and $\underline{r}_2$ are variable and the dependence on t is trivial as is the case for quasimonochromatic radiation:

$$\Gamma(\underline{r}_1,\underline{r}_2;t) = W(\underline{r}_1,\underline{r}_2)\exp(-i\omega t) \quad . \tag{2.78}$$

The spectral coherence function $W(\underline{r}_1,\underline{r}_2)$ has to be known when one wants to perform object reconstruction in partially coherent illumination (or to study fluctuating objects, see Chaps.5 and 6). We shall restrict ourselves to the phase problem of category II, the phase problem of *spatial* coherence. The phase problem of category I, the phase problem of *temporal* coherence is analogous to the phase problem for complex amplitudes which has already been amply discussed.

## 2.2.2 Determination of the Phase of the Spatial Coherence Function with an Incoherent Reference Point Source

This method, which bears a close analogy to off-axis holography [2.29], has been studied by a number of authors [2.30-32], and has been experimentally tested by BEARD [2.33], and KOHLER and MANDEL [2.34]. The quantity of ultimate interest here is the irradiance distribution across the real or effective source [2.35]. The effective source is defined as the fictitious planar source which gives rise to the same complex degree of coherence in the plane of observation, without the interven-

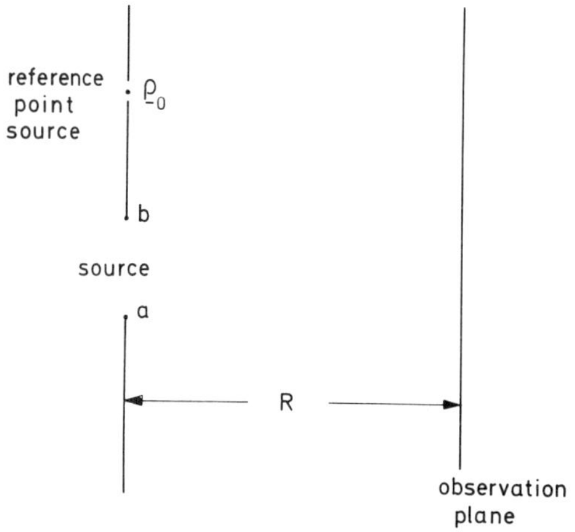


Fig. 2.3. Phase determination with incoherent point source

tion of an optical system, as the physical source with the intervention of the optical system. We shall restrict ourselves to planar sources. The extension to non-planar sources only leads to more complicated mathematics. The plane of the source and the observation plane are taken to be parallel. We consider the situation sketched in Fig.2.3. Let $s(\underline{\rho})$ be the irradiance distribution of the incoherent planar source and let an incoherent quasimonochromatic point source of strength r with the same wavelength as the source be located in the point $\underline{\rho}_0$ in the plane of the source. The total irradiance distribution $S(\underline{\rho})$ is then given by

$$S(\underline{\rho}) = r\delta(\underline{\rho}-\underline{\rho}_0) + s(\underline{\rho}) \quad . \tag{2.79}$$

According to the Van Cittert-Zernike theorem (see Sect.5.3.3) the mutual intensity in two points $\underline{r}_1$ and $\underline{r}_2$ in the plane of observation is given by:

$$W(\underline{r}_1,\underline{r}_2) = \int_\sigma \frac{S(\underline{\rho})\exp[ik(|\underline{\rho}-\underline{r}_1|-|\underline{\rho}-\underline{r}_2|)]}{|\underline{\rho}-\underline{r}_1||\underline{\rho}-\underline{r}_2|}d\underline{\rho} \quad , \tag{2.80}$$

where the integration extends over the irradiance distribution $\sigma$. In the far-field $(R\to\infty)$ (2.80) reduces to:

$$W(\underline{r}_1,\underline{r}_2) = R^{-2}\exp(i\psi) \int_\sigma S(\underline{\rho})\exp[ikR^{-1}(\underline{r}_1-\underline{r}_2)\cdot\underline{\rho}]d\underline{\rho} \quad , \tag{2.81}$$

where

$$\psi = (2R)^{-1}\ k(r_1^2-r_2^2) \quad . \tag{2.81a}$$

If we restrict our discussion to points which lie symmetrically with respect to the z-axis, which is chosen normal to the source, then $r_1 = r_2$ and (2.81) simplifies to:

$$W(\underline{r}_1,\underline{r}_2) = R^{-2} \int_\sigma S(\underline{\rho})\exp[ikR^{-1}(\underline{r}_1-\underline{r}_2)\cdot\underline{\rho}]d\underline{\rho} \quad . \tag{2.82}$$

Taking $\underline{\rho}' = (\lambda R)^{-1}\underline{\rho}$ as a new integration variable (2.82) becomes

$$W(\underline{r}_1,\underline{r}_2) = \lambda^2 \int_\sigma S(\underline{\rho}')\exp[2\pi i(\underline{r}_1-\underline{r}_2)\cdot\underline{\rho}']d\underline{\rho}' \quad . \tag{2.83}$$

Let us for simplicity take a one-dimensional situation. The measurement of the coherence function only yields $|W(x)|^2$ where x is related to $x_1$ and $x_2$ according to $x_1 = -x_2 = \frac{1}{2}x$. From (2.83) and (2.79) we straightforwardly derive:

$$\begin{aligned}\lambda^{-1}|W(x)|^2 = r^2 &+ r\int_\sigma s(\rho')\exp[2\pi ix(\rho_0'-\rho')]d\rho' + r\int_\sigma s(\rho')\exp[2\pi ix(\rho'-\rho_0')]d\rho' \\ &+ \iint_{\sigma\sigma} s(\rho')s(\rho'')\exp[2\pi ix(\rho'-\rho'')]d\rho'\,d\rho'' \quad .\end{aligned} \tag{2.84}$$

Taking the Fourier transform of this expression we find

$$\lambda^{-1}\mathrm{FT}|W(x)|^2 = r^2\delta(\rho') + rs(\rho_0'-\rho') + rs(\rho'+\rho_0') + s(\rho')^2 \quad . \tag{2.85}$$

Let us suppose that the source $s(\rho')$ extends from $\rho' = a$ to $\rho' = b$ $(b > a)$. If $\rho_0'$ is chosen so large that

$$\rho_0' > 2b - a \quad \text{or} \quad \rho_0' < b - 2a \tag{2.86}$$

then the supports of the functions $s(\rho'+\rho_0')$ and $s(\rho_0'-\rho')$ on the rhs of (2.85) are separated from the supports of the other functions and $s(\rho'-\rho_0')$ or $s^*(\rho'+\rho_0)$ can be determined directly. The decision whether $s(\rho)$ or its mirror image $s(-\rho)$ is the desired solution can only be made from the prior knowledge of the position of the point source. If $\rho_0'$ does not obey the condition (2.86) source reconstruction is still possible if

$$\rho_0' > b - a \quad . \tag{2.87}$$

In that case we have to make two measurements, one with a reference source and one without. According to (2.85), this last measurement yields $s(\rho')^2$, so that the other measurement gives $s(\rho_0'-\rho') + s(\rho'-\rho_0')$ from which $s(\rho')$ and $s(-\rho')$ can be determined because their supports do not overlap. Also in this case we need prior information to distinguish between $s(\rho')$ and $s(-\rho')$. This method has been proposed by MEHTA [2.30] and GAMO [2.31], and has been applied to astronomy by TOWNES [2.36].

### 2.2.3 Determination of the Phase of the Spatial Coherence Function with an Exponential Filter

This proposal is also due to MEHTA [2.37] and has been verified by KOHLER and MANDEL [2.34]. The method makes use of the fact that the mutual intensity is an analytic function [Ref.2.2, p.504]. Thus the logarithm of this function is analytic, except in points where $W(x) = 0$ (branch points). So $|W(z)|$ and $\phi(z) = \arg W(z)$ are connected by the Cauchy-Riemann equations:

$$\begin{aligned} \frac{\partial|W(z)|}{\partial x} &= |W(z)|\,\frac{\partial\phi(z)}{\partial y} \\ \frac{\partial|W(z)|}{\partial y} &= -|W(z)|\,\frac{\partial\phi(z)}{\partial x} \\ z &= x + iy \quad . \end{aligned} \tag{2.88}$$

If we can determine $|W(z)|$ and $(\partial/\partial y)|W(z)|$ on the real axis, we can in principle calculate $\phi(x)$ apart from an insignificant constant:

$$\phi(x) = -(\partial/\partial y) \int_0^x \ln W(z)\, dx' \Big|_{y=0} \quad . \tag{2.89}$$

In order to apply (2.89) we have to continue $W(x)$ into the complex plane. Replacing $x$ in the formula [see (2.83)],

$$W(x) = \lambda \int_\sigma S(\rho')\exp(2\pi i x\rho')d\rho' \quad , \tag{2.90}$$

by $x + iy$ we find

$$W(x+iy) = \lambda \int_\sigma S(\rho')\exp(-2\pi y\rho')\exp(2\pi i x\rho')d\rho' \quad . \tag{2.91}$$

When comparing (2.90) and (2.91) we see that $S(\rho')$ in (2.90) has been replaced by $S(\rho')\exp(-2\pi y\rho')$ in (2.91). This latter quantity may be realized in practice by placing an exponential filter with transmission function $\exp(-2\pi y\rho')$ over the source. The measurement of the visibility of the interference fringes yields $|W(x+iy)|$. This method is not easy experimentally, while also from a theoretical point of view we expect complications from possible zeros of $W(z)$ in the vicinity of the interval on the real axis on which $W(x)$ has to be determined [see the discussion in connection with (2.36)], a difficulty already noticed in [2.34].

### 2.2.4 Determination of the Phase of the Spatial Coherence Function from the Intensity in the Fraunhofer Plane

The methods described in the preceding two sections are not very practical for application to microscopy and are virtually impossible in the case of electron microscopy. With the latter application in mind FERWERDA [2.38] has proposed a method which might be useful for microscopy. The method is applicable to situations where the mutual intensity $W(\underline{r},\underline{r}')$ only depends on the difference $\underline{r} - \underline{r}'$. Let us next consider the intensity distribution in the Fraunhofer plane. This situation can be realized by locating the plane under consideration in the front focal plane of a lens. The Fraunhofer plane is then the back focal plane [Ref.2.17, p.86]. Because of the Fourier transform relationship between the complex amplitudes, the corresponding complex degrees of coherence are related according to

$$\hat{W}(\underline{\rho},\underline{\rho}') = \iint W(\underline{r}-\underline{r}')\exp[2\pi i(\underline{r}\cdot\underline{\rho}-\underline{r}'\cdot\underline{\rho}')]d\underline{r}\, d\underline{r}' \quad , \tag{2.92}$$

where $\hat{W}(\underline{\rho},\underline{\rho}')$ is the degree of coherence in the Fraunhofer plane and $W(\underline{r}-\underline{r}')$ is the degree of coherence which has to be determined. Taking again one spatial dimension and taking $\rho = \rho'$ we find from (2.92) the result

$$I(\rho) = \int_0^{2a} dx(2a-x)[W(-x)\exp(-2\pi i\rho x)+W(x)\exp(2\pi i\rho x)] \quad , \tag{2.93}$$

where $I(\rho)$ is the intensity in the point $\rho$ of the Fraunhofer plane. It has been assumed that the x-interval is $-a \leqq x \leqq a$ corresponding to an aperture of width 2a centered on the optic axis. Taking the Fourier transform of $I(\rho)$, viz.

$$\hat{i}(x) = \int_{-\infty}^{\infty} I(\rho)\exp(2\pi i\rho x)d\rho \quad , \tag{2.94}$$

we find after an easy calculation

$$W(x) = \begin{cases} (2a-x)^{-1}\hat{i}(-x) & \text{for } 0 \leqq x \leqq 2a \\ (2a+x)^{-1}\hat{i}(-x) & \text{for } -2a < x \leqq 0 \end{cases} \quad . \tag{2.95}$$

Experiments should decide upon the practicality of this proposal. The corresponding radiometric problem is considered in Section 5.4.

## References


2.1 A. Walther: Opt. Acta *10*, 41 (1963)
2.2 M. Born, E. Wolf: *Principles of Optics*, 4th ed. (Pergamon Press, London, New York 1970) Chap.9
2.3 B.J. Hoenders: On the inversion of an integral equation relating two wave functions in planes of an optical system suffering from an arbitrary number of aberrations. Opt. Acta (to be published)
H.A. Ferwerda, B.J. Hoenders: Optik *39*, 317 (1974)
2.4 E.T. Copson: *Theory of Functions of a Complex Variable* (Oxford University Press, Oxford 1950) pp.107, 108
2.5 B.J. Hoenders: J. Math. Phys. *16*, 1719 (1975)
2.6 E.M. Hofstetter: IEEE Trans. Inf. Theory *10*, 119 (1964)
2.7 A.H. Greenaway: Opt. Lett. *1*, 10 (1977)
2.8 J. Hilgevoord: *Dispersion Relations and Causal Description* (North-Holland, Amsterdam 1960)
2.9 D.L. Misell, R.E. Burge, A.H. Greenaway: J. Phys. D *7*, L27 (1974);
D.L. Misell, A.H. Greenaway: J. Phys. D *7*, 832 (1974);
R.E. Burge, M.A. Fiddy, A.H. Greenaway, G. Ross: J. Phys. D *7*, 65 (1974)
2.10 R.E. Burge, M.A. Fiddy, A.H. Greenaway, G. Ross: Proc. Roy. Soc. London A *350*, 191 (1976)
2.11 E. Wolf: J. Opt. Soc. Am. *60*, 18 (1970)
2.12 E.C. Titchmarsh: Proc. Lond. Math. Soc.(2) *25*, 283 (1926) Lemma 4.4
2.13 D.L. Misell: J. Phys. D *6*, L6, 2200, 2217 (1973). See comment by R.W. Gerchberg, W.O. Saxton: J. Phys. D *6*, L31 (1973)
2.14 A.J.J. Drenth, A.M.J. Huiser, H.A. Ferwerda: Opt. Acta *22*, 615 (1975)
2.15 A.M.J. Huiser, H.A. Ferwerda: Opt. Acta *23*, 445 (1976)
2.16 P. van Toorn, H.A. Ferwerda: Opt. Acta *23*, 457 (1976)
2.17 J.W. Goodman: *Fourier Optics* (McGraw-Hill, New York 1968) p.25
2.18 R.W. Gerchberg, W.O. Saxton: Optik *34*, 275 (1971)
2.19 R.W. Gerchberg, W.O. Saxton: Optik *35*, 237 (1972)
2.20 A.M.J. Huiser, A.J.J. Drenth, H.A. Ferwerda: Optik *45*, 303 (1976)
2.21 A.M.J. Huiser, H.A. Ferwerda: Optik *46*, 407 (1976)
2.22 P. Schiske: Optik *40*, 261 (1974)
2.23 A.M.J. Huiser, P. van Toorn, H.A. Ferwerda: Optik *47*, 1 (1977);
P. van Toorn, H.A. Ferwerda: Optik *47*, 123 (1977)

2.24 W.J. Dallas: Optik *41*, 45 (1975)
2.25 P. van Toorn, A.M.J. Huiser, H.A. Ferwerda: To be published
2.26 E. Wolf: Proc. Phys. Soc. London *80*, 1269 (1962)
2.27 D. Dialetis: J. Math. Phys. *8*, 1641 (1967);
D. Dialetis, E. Wolf: Nuovo Cimento *47*, 113 (1967)
2.28 H.M. Nussenzveig: J. Math. Phys. *8*, 561 (1967)
2.29 E.N. Leith, J. Upatnieks: J. Opt. Soc. Am. *53*, 1377 (1963)
2.30 C.L. Mehta: J. Opt. Soc. Am. *58*, 1233 (1968)
2.31 H. Gamo: In *Electromagnetic Theory and Antennas*, ed. by E.C. Jordan (Pergamon Press, Oxford 1963) p.801
2.32 J.W. Goodman: J. Opt. Soc. Am. *60*, 506 (1970)
2.33 T.D. Beard: J. Opt. Soc. Am. *59*, 1525 A (1969); Appl. Phys. Lett. *15*, 227 (1969)
2.34 D. Kohler, L. Mandel: J. Opt. Soc. Am. *60*, 280 (1970); *63*, 126 (1973)
2.35 H.H. Hopkins: Proc. Roy. Soc. A *208*, 263 (1951); A *217*, 408 (1953)
2.36 C.H. Townes: Phys. Today *25* (7), 17 (1972)
2.37 C.L. Mehta: Nuovo Cimento *36*, 202 (1965)
2.38 H.A. Ferwerda: Opt. Commun. *19*, 54 (1976)

# 3. The Uniqueness of Inverse Problems

B. J. Hoenders

With 2 Figures

The retrieval of the phase of the scattered field as studied in Chapter 2 is the first step in the determination of the structure of the scattering objects. The next step of reconstructing the field up to the surface of the scatterer, and the final step of reconstructing the object from the field outside the object are reviewed in the present chapter with emphasis on the uniqueness of the reconstruction.

## 3.1 Summary of Inverse Problems

In physics frequently the problem arises whether or not certain properties of a field (object) or the field (object) itself can be determined from an observable quantity. A famous example of such an inverse problem is the determination of the structure of a crystal from its x-ray diffraction pattern (HOSEMANN and BAGHI [3.1]). Quite another inverse problem was solved as early as 1933 by LANGER [3.2], who showed that if an electrode is placed on top of an infinite plane which is the boundary of a medium whose conductivity only depends on the direction normal to the surface, the conductivity is uniquely determined by the values of the potential at the surface.

### 3.1.1 Inverse Sturm-Liouville Problems

Another very important example of an inverse problem is the inverse Sturm-Liouville (S-L) problem, i.e., the determination of a function q from the knowledge of the spectrum (or spectra) of the eigenvalue problem $(L+\lambda^2+q)\psi = 0$, where L denotes a more dimensional linear operator, and the function $\psi$ satisfies a boundary condition. A related problem is formulated by KAC [3.3], (can one hear the shape of a drum). Similar problems are formulated by BALTES and HILF [3.4], (can one see the shape of a blackbody).

Another physical example of an inverse S-L problem is the determination of the index of refraction of a sphere from the knowledge of its natural frequencies, (BORN and WOLF [Ref.3.5, Sect.13.5.3], DEBEYE [3.6]) or, in general, the determination of the tensors $\underline{\underline{\varepsilon}}(\underline{r};k)$, $\underline{\underline{\mu}}(\underline{r};k)$, and $\underline{\underline{\sigma}}(\underline{r};k)$ from the spectrum of natural oscillations of an object (see also [3.87]).

The inverse S-L problem is not further considered in this chapter. We refer to the paper by BORG [3.7], who showed that q is uniquely determined from the knowledge of two spectra generated by two different homogeneous boundary conditions, and to an earlier paper by AMBARZUMIAN [3.8], who obtained a similar result with a rather elementary calculation. Both results, obtained for the one-dimensional problem, are very likely to be applicable to more dimensions. Surveys of inverse problems are given by NEWTON [3.9] and BALTES [3.10].

### 3.1.2 Reconstruction Problems

The reconstruction of scattering objects from measurements of the scattered field can be divided into three separate problems:

a) From a measurable quantity like the intensity or the differential cross section, the phase of the field on a surface is to be determined (Phase problem).

b) From the knowledge of the field on a surface, the field up to the scatterer is to be reconstructed (Inverse diffraction problem).

c) The object is to be reconstructed from the knowledge of the field outside the object (Inverse source problem).

Problem a) has been analyzed in Chapter 2 of this book. Problem b) is analyzed in Section 3.2, where it is shown that a vector field $\underline{A}$, satisfying $\nabla\times\nabla\times\underline{A}-k^2\underline{A}=0$, and a scalar field $\psi$, satisfying $(\nabla^2+k^2)\psi=0$, such that both $\underline{A}$ and $\psi$ satisfy the radiation condition, can be explicitly calculated from their far-field patterns on any spherical surface enclosing the object.

In Section 3.2.2 procedures are derived by which the values of the fields $\underline{A}$ and $\psi$ on any spherical surface enclosing the object can be determined from their values on any spherical surface to which the field has propagated. It is shown in Section 3.2.4 that the values of a scalar field on any surface enclosing the scatterer are uniquely determined by the values of $\psi$ on any closed surface to which the field has propagated, provided that $\psi$ satisfies Sommerfeld's radiation condition at infinity. It is shown in Section 3.2.5 that, relying on the Kirchhoff approximation or the geometrical optical approximation, the shape of a perfect conductor sometimes can be determined from the scattered field or the differential cross section.

Problem c) is analyzed in the remaining sections. Unfortunately, it frequently occurs that the knowledge of a field outside an object is not sufficient to determine the object uniquely. Several examples of a nonunique relation between an object and scattering data are constructed in Section 3.4.1. For example it is shown that for every incoming wave an infinite number of potentials with finite support can be constructed leading to a vanishing scattered field outside the support of the potential. It is even possible to construct a physically realizable potential (or equivalently a scalar index of refraction) such that any member of any finite set of incoming monochromatic plane waves with different wave vectors leads to a vanishing scattered field outside the object. Moreover, it is shown that many charge-current distributions are not radiating, and therefore necessarily lead to at most a static field outside the distribution. For example, when the center of a uniformly charged spherical shell with total charge e and radius a in purely translatory motion describes a closed orbit periodically in a time $2a(cn)^{-1}$, where n denotes an integer, and c the velocity of light, the electromagnetic field in every outer point is purely static (SCHOTT [3.11], ARNETT and GOEDECKE [3.12]). The theory of nonradiating distributions is developed on using the multipole expansions of the electromagnetic field (DEVANEY and WOLF [3.13], see Sect.3.3.2) or the theory of integral equations (COHEN and BLEISTEIN [3.14], see Sect.3.3.3).

However, uniqueness may be obtained using additional information. For example, it is shown that an object, characterized by either a potential or index of refraction, both with finite support, is uniquely determined by the scattered fields generated by either an infinite set of monochromatic plane waves with different wave vectors, or an infinite set of nonmonochromatic waves (Sect.3.4.4). Prior knowledge about the object may also lead to the desired uniqueness, or restrict the class of solutions (Sect.3.4.2). MIRELESS [3.15] has shown that the constant index of refraction of an infinite cylinder is uniquely determined by the scattered field generated by an incoming linearly polarized monochromatic plane wave. In general, the knowledge of the scattered field generated by a single plane wave is not sufficient to determine an object uniquely. However, SCHMIDT-WEINMAR et al. [3.16] have shown that a unique reconstruction of the object is possible provided that it is known a priori that the scattered field can be described in first-order Born approximation and that the object can be divided into blocklets whose order of magnitude is of the order of a wavelength (for a criticism, see [3.88]).

### 3.1.3 Three-Dimensional Reconstruction from Projections

If an object is illuminated by a plane wave and the rays through the object are approximately straight lines, a projection of the object is measured. The question whether or not an object can be reconstructed from its projections is not a standard part of 19th century mathematics, as suggested by CORMACK [3.17], but rather of 20th century mathematics, and was solved for a two-dimensional function by RADON [3.18] in 1917. The general case was solved by MADER [3.19], who has shown the connection of this problem with an initial value problem for an n-dimensional hyperbolic operator. A simple proof has been obtained by UHLENBECK [3.20] (see also GEL'FAND and SCHILOW [Ref.3.21, Vol.1, Chap.1, Sect.3.11]). This theory has become very important in electron microscopy. Although this problem is not considered in the present chapter, we will for the sake of completeness give a few references. ZWICK and ZEITLER [3.22] give an excellent review of the problem together with many references, and show the connection of several methods to determine a function from its projections. We mention also the papers by CORMACK [3.17], MARR [3.23], and ZEITLER [3.24].

The problem of the stability of a particular reconstruction procedure is not considered in this chapter. We refer to the extensive survey paper by TURCHIN et al. [3.25], and the book by LAVRENTIEV [3.26].

## 3.2 Inverse Diffraction

Suppose a scalar field $\psi(\underline{r})$ satisfying

$$(\nabla^2+k^2)\psi = 0 \quad , \tag{3.1}$$

in a domain which does not contain an object is a superposition of an incoming wave $\psi^{inc}(\underline{r})$ and a scattered wave $\psi^{sc}(\underline{r})$. The scattering is, for instance, due to a potential distribution or a medium characterized by an index of refraction, both with finite support. Then the problem arises whether or not the field between the scatterer and a surface S lying outside the scatterer can be determined from the values of the field on S. The analogous electromagnetic problem would be the reconstruction of a vector field $\underline{A}$, satisfying

$$\nabla \times \nabla \times \underline{A} - k^2\underline{A} = 0 \tag{3.2}$$

outside the scatterer from its values on the surface S. Two constructive methods have been developed to analyze this problem, showing that for certain geometries a unique determination of the field, up to the scatterer, is possible.

In Section 3.2.1 a procedure is formulated which, for the scalar case can be traced back to SOMMERFELD [3.27], and in the vectorial case to WILCOX [3.28]. This procedure shows that the far-field pattern suffices to determine the field in any spherical surface enclosing the scatterer. In Sections 3.2.2 and 3.2.3 we formulate procedures with which the field on a plane or a spherical surface can be calculated from its values on a plane or a spherical surface to which the field has propagated.

### 3.2.1 Inverse Diffraction from Far-Field Data

Let $\psi(\underline{r})$ be a solution of

$$(\nabla^2+k^2)\psi(\underline{r}) = 0 \quad , \tag{3.3}$$

outside a finite domain D, satisfying Sommerfeld's radiation condition at infinity.

On expanding $\psi(\underline{r})$ on the surface $r=a$ into a set of surface spherical harmonics $Y_\ell^m(\theta,\phi)$ we obtain

$$\psi(r,\theta,\phi) = \sum_{\ell,m} C_{\ell,m}\, Y_{\ell,m}(\theta,\phi) h_\ell^{(1)}(kr) \quad , \tag{3.4}$$

if

$$C_{\ell,m} = \int_\Omega d\Omega\ \psi(a,\theta,\phi) Y_{\ell,m}(\theta,\phi) [h_\ell^{(1)}(ka)]^{-1} \tag{3.5}$$

and $h^{(1)}(kr)$ denotes the spherical Bessel function of the first kind. Using

$$h_\ell^{(1)}(kr) = \frac{\exp(ikr)}{kr} \sum_{j=0}^{\ell} (\ell,j)(2irk)^{-j} \quad , \tag{3.6}$$

where

$$(\ell,j) = 2^{-2j}(j!)^{-1}(4\ell^2-1)(4\ell^2-9)\ldots[4\ell^2-(2j-1)^2] \quad , \quad (\ell,0) = 1 \quad , \tag{3.7}$$

denotes Hankel's symbol, combination of (3.4-6) leads to:

$$\psi(r,\theta,\phi) = \frac{\exp(ikr)}{kr} \sum_{\ell=0}^{\infty} f_\ell(\theta,\phi) r^{-\ell} \quad , \tag{3.8}$$

if

$$f_\ell(\theta,\phi) = \sum_{\ell'=j}^{\infty} \sum_{m=-\ell'}^{+\ell'} (\ell',\ell) C_{\ell',m}\, Y_{\ell',m}(\theta,\phi)(2ik)^{-\ell} \quad . \tag{3.9}$$

The functions $f_\ell$ satisfy a simple recursion formula. Inserting (3.8) into (3.3) yields:

$$\frac{\exp(ikr)}{kr} \sum_{\ell=0}^{\infty} \left[-r^{-\ell-1}(2ik\ell)+r^{-\ell-2}\ell(\ell+1)+r^{-\ell-2}D\right] f_\ell = 0 \quad , \tag{3.10}$$

where

$$D = (\sin\theta)^{-1} \frac{\partial}{\partial\theta}(\sin\theta \frac{\partial}{\partial\theta}) + (\sin\theta)^{-2} \frac{\partial^2}{\partial\phi^2} \quad . \tag{3.11}$$

Equating the coefficients of $r^{-\ell-2}$, $\ell = 0,\ldots$ leads to

$$2ik(\ell+1) f_{\ell+1} = [\ell(\ell+1)+D] f_\ell \quad . \tag{3.12}$$

However, (3.12) shows that all the functions $f_\ell$ can be determined from the knowledge of the far-field amplitude $\exp(ikr/r)f_0(\theta,\phi)$, and that therefore the field $\psi$ can be determined outside every spherical surface enclosing the scatterer from its far-field pattern.

A similar result has been obtained for a vector field by WILCOX [3.28]. His analysis is based on a representation theorem for vector fields satisfying

$$\nabla \times \nabla \times \underline{A} - k^2\underline{A} = 0 \quad , \tag{3.13}$$

together with the vectorial form of the radiation condition

$$\lim_{r\to\infty} r\{\hat{\underline{r}}\times(\nabla\times\underline{A})+ik\underline{A}\} = 0 \quad , \tag{3.14}$$

where $\hat{\underline{r}}$ denotes the unit radial vector. We will need the additional condition

$$\underline{A}(\underline{r}) = O\{r^{-1}\} \quad , \quad \text{if } r \to \infty \quad . \tag{3.15}$$

The desired representation theorem is derived from a vectorial form of Green's theorem. From the divergence theorem,

$$\int_\tau \nabla \cdot \underline{F}d\tau = \int_\sigma \underline{n} \cdot \underline{F}d\sigma \quad , \tag{3.16}$$

where $\underline{n}$ is a unit vector normal to the boundary surface $\sigma$ of a volume $\tau$, pointing out of $\sigma$, applied to the vector

$$\underline{F} = \underline{A} \times (\nabla\times\underline{B}) \quad , \tag{3.17}$$

and using the identity

$$\nabla \cdot (\underline{A}\times\nabla\times\underline{B}) = (\nabla\times\underline{A}) \cdot (\nabla\times\underline{B}) - \underline{A} \cdot \nabla \times (\nabla\times\underline{B}) \quad , \tag{3.18}$$

we obtain

$$\int_\tau [\underline{A}\cdot\nabla\times(\nabla\times\underline{B})-(\nabla\times\underline{A})\cdot(\nabla\times\underline{B})]d\tau = \int_\sigma (\nabla\times\underline{B}) \times \underline{A} \cdot \underline{n}\, d\sigma \quad . \tag{3.19}$$

Interchanging $\underline{A}$ and $\underline{B}$ in (3.19) and subtracting the result from (3.19) yields

$$\int_\tau [A\circ\nabla\times(\nabla\times\underline{B})-\underline{B}\cdot\nabla\times(\nabla\times\underline{A})]d\tau = \int_\sigma [(\nabla\times\underline{B})\times\underline{A}-(\nabla\times\underline{A})\times B] \cdot \underline{n}\, d\sigma \quad . \tag{3.20}$$

Application of (3.20) to the field $\underline{A}$ satisfying (3.14) and (3.15) and to the field

$$\underline{B} = \underline{u}\,\frac{\exp ik|\underline{r}-\underline{r}'|}{4\pi|\underline{r}-\underline{r}'|} \quad , \tag{3.21}$$

where $\underline{u}$ is a constant vector, in the region between a large sphere $S_R$ with radius R enclosing a surface S and a sphere $S_\alpha$ of radius $\alpha$ both with center at the vector $\underline{r}$ outside S together with the relations

$$\nabla\times\nabla\times\underline{B} - k^2\underline{B} = \nabla\cdot(\underline{u}\cdot\nabla)\,\frac{\exp ik|\underline{r}-\underline{r}'|}{4\pi|\underline{r}-\underline{r}'|} \tag{3.22}$$

and

$$\underline{A}\cdot\nabla(\underline{u}\cdot\nabla)\,\frac{\exp ik|\underline{r}-\underline{r}'|}{|\underline{r}-\underline{r}'|} = \nabla\left[\underline{A}(\underline{u}\cdot\nabla)\,\frac{\exp ik|\underline{r}-\underline{r}'|}{|\underline{r}-\underline{r}'|}\right] \tag{3.23}$$

yields

$$\int\limits_{S+S_\alpha+S_R} \{G(\underline{r},\underline{r}';k)\underline{u}\times(\nabla\times\underline{A})\cdot\underline{n}-\underline{A}\times[\nabla G(\underline{r},\underline{r}';k)\times\underline{u}]$$
$$\cdot\underline{n}-[\underline{u}\cdot\nabla G(\underline{r},\underline{r}';k)]\times(\underline{A}\cdot\underline{n})\}d\sigma = 0 \tag{3.24}$$

with

$$G(\underline{r},\underline{r}';k) = \frac{\exp ik|\underline{r}-\underline{r}'|}{4\pi|\underline{r}-\underline{r}'|} \quad . \tag{3.25}$$

A straightforward calculation shows

$$\lim_{\alpha\to\infty}\int\limits_{S_\alpha} \ldots = -\underline{A}(\underline{r})\cdot\underline{u} \tag{3.26}$$

and

$$\int\limits_{S_R} \ldots = \frac{1}{4\pi}\int\limits_{\Omega}\exp(ikR)R\{-u\cdot\hat{\underline{r}}\times(\nabla\times\underline{A})+(ik-R^{-1})\underline{A}\cdot[\hat{\underline{r}}\times(\hat{\underline{r}}\times\underline{u})]$$
$$-(ik-R^{-1})(\underline{u}\cdot\hat{\underline{r}})(\underline{A}\cdot\hat{\underline{r}})\}d\Omega = \frac{1}{4\pi}\,u\cdot\left\{\int\limits_{\Omega}\exp(ikR)R[\hat{\underline{r}}\times(\nabla\times\underline{A})+ik\underline{A}]d\Omega\right.$$
$$\left.-\int\limits_{\Omega}\exp(ikR)\underline{A}d\Omega\right\} \quad . \tag{3.27}$$

In view of the relations (3.14) and (3.15), these expressions tend to zero if $R\to\infty$. This result has been derived by WILCOX [3.28] without the assumption (3.15), but with the weaker assumption

$$\lim_{R\to\infty} \int_{r=R} |\hat{\underline{r}}\times(\nabla\times\underline{A})+ik\underline{A}|^2 d\sigma = 0 \quad . \tag{3.28}$$

Combination of (3.24,26,27) leads to the desired representation theorem in the limit $R\to\infty$:

$$\underline{A}(\underline{r}) = \int_\sigma [\,\underline{n}\times(\nabla\times\underline{A})G(\underline{r},\underline{r}';k)+(\underline{n}\times\underline{A})\times\nabla G(\underline{r},\underline{r}';k)+(\underline{A}\cdot\underline{n})\nabla G(\underline{r},\underline{r}';k)]d\sigma \quad . \tag{3.29}$$

From (3.29) we derive the analog of (3.8),

$$\underline{A}(\underline{r}) = \frac{\exp(ikr)}{r}\sum_{n=0}^{\infty} r^{-n}\,\underline{A}_n(\theta,\phi) \quad , \tag{3.30}$$

valid for all values of r larger than the radius c of a sphere enclosing the scatterer. The expansion follows immediately from the observation that $G(\underline{r},\underline{r}';k)$ admits a power series expansion into inverse powers of $|\underline{r}|$ for all values of $|\underline{r}| > c$.

A recursion formula for the functions $\underline{A}_n$ has been established by WILCOX [3.28]. We will omit the rather lengthy proof and give Wilcox's result.

The coefficients $\underline{A}_n$ in the expansion are determined by the radiation pattern $\underline{A}_0(\theta,\phi)$ satisfying

$$\underline{A}(\underline{r}) \sim \frac{\exp(ikr)}{r}\,\underline{A}_0(\theta,\phi) \tag{3.31}$$

where

$$\underline{A}_0\cdot\underline{r} = 0 \quad , \text{or} \quad \underline{A}_0 = A_0^2\underline{\theta} + A_0^3\underline{\phi} \quad , \tag{3.32}$$

through the recursion formulas

$$ikA_1^1 = -(\sin\theta)^{-1}\left[\frac{\partial}{\partial\theta}(\sin\theta\, A_0^2) + \frac{\partial A_0^3}{\partial\phi}\right] = -\underline{r}\cdot\nabla\underline{A}_0 \quad , \tag{3.33}$$

$$2ikn\, A_{n+1}^1 = n(n-1)A_n^1 + D\, A_n^1 \quad , \quad n = 1,2,3,\ldots \quad , \tag{3.34}$$

and

$$2ikn\, A_n^2 = n(n-1)A_{n-1}^2 + D\, A_{n-1}^2 + D_\theta A_{n-1} \quad , \quad n = 1,2,3,\ldots \quad , \tag{3.35a}$$

$$2ikn\, A_n^3 = n(n-1)A_{n-1}^3 + D\, A_{n-1}^3 + D_\phi A_{n-1} \quad , \quad n = 1,2,3,\ldots \quad , \tag{3.35b}$$

where the operator D is defined by (3.11) and is known as Beltrami's operator for the sphere, while $D_\theta$ and $D_\phi$ are first-order linear operators (taking vector fields $F=F^1\underline{r}+F^2\underline{\theta}+F^3\underline{\phi}$ into scalar functions) defined by

$$D_\theta F = 2\,\frac{\partial F^1}{\partial\theta} - \frac{1}{\sin^2\theta}\,F^2 - \frac{2\cos\theta}{\sin^2\theta}\,\frac{\partial F^3}{\partial\phi} \quad , \tag{3.36}$$

and

$$D_\phi F = \frac{2}{\sin\theta}\,\frac{\partial F^1}{\partial\phi} + \frac{2\cos\theta}{\sin^2\theta}\,\frac{\partial F^2}{\partial\phi} - \frac{1}{\sin^2\theta}\,F^3 \quad . \tag{3.37}$$

We now consider the problems concerning the determination of the values of a field at a sphere with radius a from its values at a sphere with radius b, $b > a$ (Sect. 3.2.2), and the determination of a field in a plane $z = 0$ from its values at a plane $z = b$ to which the field has propagated (Sect.3.2.3).

### 3.2.2 Inverse Diffraction from Spherical Surface to Spherical Surface

For the discussion of the scalar case we consider the expansion (3.3), viz.,

$$\psi(r,\theta,\phi) = \sum_{\ell=0} \sum_{m=-\ell}^{+\ell} C_{\ell,m}\, h_\ell^{(1)}(kr) Y_\ell^m(\theta,\phi) \quad , \tag{3.38}$$

and

$$C_{\ell,m} = \left[h_\ell^{(1)}(kr)\right]^{-1} \int_\Omega d\Omega\psi(r,\theta,\phi) Y_\ell^{m*}(\theta,\phi) \quad . \tag{3.39}$$

Equation (3.38) shows that if $\psi$ in $r = a$ is expanded into the complete set of functions $Y_\ell^m$ with expansion coefficients $C_{\ell,m}$ and similarly at $r = b$ with expansion coefficients $C_{\ell,m}^{(1)}$ that

$$C_{\ell,m}^{(1)} = C_{\ell,m}\,\frac{h_\ell^{(1)}(kb)}{h_\ell^{(1)}(ka)} \quad . \tag{3.40}$$

Hence, the values of $\psi(\underline{r})$ on the surface $r = a$ can be calculated from its values at $r = b$ by

$$\psi(a,\theta,\phi) = \sum_{\ell=0}^{\infty} \sum_{m=-\ell}^{+\ell} \frac{h_\ell^{(1)}(ka)}{h_\ell^{(1)}(kb)} \int_\Omega d\Omega\psi(b,\theta',\phi') Y_\ell^{m*}(\theta',\phi') Y_\ell^m(\theta,\phi) \quad . \tag{3.41}$$

A vector field $\underline{A}$, satisfying (3.2) can be expanded into two sets of vector spherical harmonics (DEVANEY and WOLF [3.29], BLATT and WEISSKOPF [3.30]), viz.,

$$\underline{A}(\underline{r}) = \sum_{\ell=1}^{\infty} \sum_{m=-\ell}^{+\ell} \left[a_\ell^m \underline{E}_{\ell,m}^e(\underline{r}) + b_\ell^m \underline{E}_{\ell,m}^h(\underline{r})\right] \quad , \tag{3.42}$$

where

$$\underline{E}^h_{\ell,m}(\underline{r}) = ik\,\nabla \times \left[\hat{\underline{r}} h^{(1)}_\ell(kr) Y^m_\ell(\theta,\phi)\right] \quad , \tag{3.43}$$

and

$$\underline{E}^e_{\ell,m}(\underline{r}) = \nabla \times \left\{\nabla\times\left[\hat{\underline{r}} h^{(1)}_\ell(kr) Y^m_\ell(\theta,\phi)\right]\right\} \quad . \tag{3.44}$$

The vector spherical harmonics $\underline{E}^e_{\ell,m}$ and $\underline{E}^h_{\ell,m}$ are orthogonal in the surface $r=a$ (JONES [Ref.3.31, Sect.8.17]); hence functions P(a) and Q(a) exist such that

$$\int_\Omega d\Omega\, \underline{E}^h_{\ell,m}(\underline{r}) \cdot \underline{E}^{h*}_{s,t}(\underline{r}) = \delta_{\ell,s}\,\delta_{m,t}\left\{-k^2\ell(\ell+1)[h^1_\ell(ka)]^2\right\} = \delta_{\ell,s}\,\delta_{m,t}\,P(a) \quad , \tag{3.45}$$

$$\int_\Omega d\Omega\, \underline{E}^e_{\ell,m}(\underline{r}) \cdot \underline{E}^{e*}_{s,t}(\underline{r}) = \delta_{\ell,s}\,\delta_{m,t}\, a^{-2}\ell(\ell+1)\Big(\ell(\ell+1)[h^1(ka)]^2 + \left\{\frac{d}{da}[ah^1_\ell(ka)]\right\}^2\Big) = \delta_{\ell,s}\,\delta_{m,t}\,Q_\ell(a) \quad . \tag{3.46}$$

Moreover,

$$\int_\Omega d\Omega\, \underline{E}^e_{\ell,m}(\underline{r}) \cdot \underline{E}^h_{s,t}(\underline{r}) = 0 \quad . \tag{3.47}$$

Equations (3.42) to (3.47) lead to

$$a_{\ell,m} = \int_\Omega d\Omega\, \underline{A}(\underline{r}) \cdot \frac{\underline{E}^{e*}_{\ell,m}(\underline{r})}{Q_\ell(b)} \quad , \tag{3.48}$$

and

$$b_{\ell,m} = \int_\Omega d\Omega\, \underline{A}(\underline{r}) \cdot \frac{\underline{E}^{h*}_{\ell,m}(\underline{r})}{P_\ell(b)} \quad , \tag{3.49}$$

if the field is expanded on the surface $r=b$.

From the values of the expansion coefficients, determined by (3.48) and (3.49), we can determine the value of $\underline{A}$ in a spherical surface $r=a$ using the same procedure as developed for the scalar case. To be more specific, decompose the field (3.42) into the various components of the vector spherical wave functions, which are of the form $a^m_\ell g_\ell(b)\Phi^m_\ell(\theta,\phi)$ or $b^m_\ell g_\ell(b)\Phi^m_\ell(\theta,\phi)$, i.e., a coefficient $a^m_\ell$ or $b^m_\ell$ times the product of a function of $r=b$ and a function of $\theta$ and $\phi$. The corresponding quantity on the surface $r=a$ is found multiplying the known coefficients in front of $\Phi^m_\ell(\theta,\phi)$ by $g_\ell(a)/g_\ell(b)$.

### 3.2.3 Inverse Diffraction from Plane to Plane

The calculations of this section are based on methods developed by SHEWELL and WOLF [3.32] and SHERMAN [3.33]. Let $\hat{\psi}(a,f_x,f_y)$ denote the spatial Fourier transform of the wave function $\psi(\underline{r})$ in the plane $z=a$

$$\hat{\psi}(a,f_x,f_y) = \int_{-\infty}^{+\infty} dx \int_{-\infty}^{+\infty} d_y \exp[-2\pi i(f_x x+f_y y)]\psi(a,x,y) \quad . \tag{3.50}$$

Then the wave function at the plane $z=b$ reads

$$\psi(b,x,y) = \int_{-\infty}^{+\infty} df_x \int_{-\infty}^{+\infty} df_y \exp\left\{2\pi i[f_x x+f_y y+m(b-a)]\right\}\hat{\psi}(a,f_x,f_y) \quad , \tag{3.51}$$

if

$$m = (k^2-f_x^2-f_y^2)^{1/2} \quad , \quad \mathrm{Im}\left\{(k^2-f_x^2-f_y^2)^{1/2}\right\} > 0 \quad , \text{ and } \quad b \geqq a \cdot , \tag{3.52}$$

because each wave $\exp 2\pi i(f_x x+f_y y+mz)$ is a solution of the Helmholtz equation. Therefore, the spatial Fourier transform of the wave function in the plane $z=b$, $\hat{\psi}(b,f_x,f_y)$ is related to $\hat{\psi}(a,f_x,f_y)$ by

$$\hat{\psi}(b,f_x,f_y) = \exp\{2\pi im(b-a)\}\hat{\psi}(a,f_x,f_y) \quad . \tag{3.53}$$

Combination of (3.50) and (3.53) leads to

$$\psi(a,x,y) = \int_{-\infty}^{+\infty} df_x \int_{-\infty}^{+\infty} df_y \exp\left\{2\pi i[f_x x+f_y y-m(b-a)]\right\}\hat{\psi}(b,f_x,f_y) \quad . \tag{3.54}$$

Some care should be exercised with the integrals and summations occuring in (3.41) and (3.54). For instance, (3.53) shows that the high frequency components (by which we mean those components for which $f_x^2+f_y^2>k^2$) are attenuated in the propagation process. They have therefore to be amplified in the restoration process. This is achieved by dividing every spectral component by $\exp 2\pi im(b-a)$. It is for this reason that the kernel of the integral (3.54) becomes *unbounded*. One therefore expects that a reliable restoration is only possible for values of $f_x$ and $f_y$ such that $f_x^2+f_y^2<k^2$. This leads to a better understanding of the meaning of the concept of the number of degrees of freedom of a wave field (HOENDERS and FERWERDA [3.34], MIYAMOTO [3.35]). A similar discussion can be set up for the reconstruction of the values of a scalar field in a spherical surface, (3.41), using the result that for large values of $\ell$ one has

$$h_\ell^{(1)}(x) = \frac{2^n i(n-1)!}{\pi^{1/2}\, x^{n+1/2}\, 2^{1/2}} + O\left\{\frac{1}{\ell}\right\} \quad . \tag{3.55}$$

The asymptotic expansion (3.55) is easily deduced from the representation of $h_\ell^{(1)}(x)$ by a Laurent series plus a logarithmic term, and therefore

$$\frac{h_\ell^{(1)}(ka)}{h_\ell^{(1)}(kb)} \simeq \left(\frac{b}{a}\right)^\ell \quad . \tag{3.56}$$

## 3.2.4 Generalization to Arbitrary Surfaces

The geometries of the sphere and the plane make an explicit calculation of the inverse diffraction problem feasible because the Helmholtz equation is separable for these geometries. However, the question arises if in principle a procedure can be developed with which the inverse diffraction problem can be solved for arbitrary geometries. This problem is solved in this section.

Suppose that $\psi(\underline{r})$ satisfies the Helmholtz equation outside a surface $\sigma$ and

$$\psi(\underline{r}) = f(\underline{r}) \quad \text{if} \quad \underline{r} \in \sigma \quad . \tag{3.57}$$

If $\psi$ satisfies Sommerfeld's radiation condition, there exists a unique Green's function $H(\underline{r},\underline{r}';k)$ such that

$$\psi(\underline{r}) = \int_\sigma H(\underline{r},\underline{r}';k) f(\underline{r}')d\sigma \quad . \tag{3.58}$$

Let $f(\underline{r})$ be developed in a complete orthonormal set of functions $\{\phi_n(\underline{r})\}$,

$$f(\underline{r}) = \sum_n a_n \phi_n(\underline{r}) \quad , \tag{3.59}$$

with

$$a_n = \int_\sigma f(\underline{r}')\phi_n(\underline{r}')d\sigma' \quad . \tag{3.60}$$

Inserting (3.59) into (3.58) leads to

$$\psi(\underline{r}) = \sum_n a_n \psi_n(\underline{r};k) \quad , \tag{3.61}$$

where

$$\psi_n(\underline{r};k) = \int_\sigma H(\underline{r},\underline{r}';k)\phi_n(\underline{r}')d\sigma \quad . \tag{3.62}$$

The problem is whether or not the coefficients $a_n$ can be uniquely determined from (3.61) if $\underline{r}$ belongs to a surface S enclosing $\sigma$. The numbers $a_n$ are uniquely determined if the set of functions $\{\psi_n(\underline{r};k), \underline{r} \in S\}$ is linearly independent. However, if this set were linearly dependent there would exist numbers $b_j$ such that

$$\sum_j b_j \psi_j(\underline{r};k) = 0 \quad \text{if} \quad \underline{r} \in S \quad , \tag{3.63}$$

where the left-hand side of (3.63) is a solution of the Helmholtz equation if $\underline{r}$ is situated outside S which satisfies Sommerfeld's radiation condition at infinity. However, the only solution of the Helmholtz equation which is zero on S and satisfie Sommerfeld's radiation condition is identically zero outside S. We recall that every solution of an elliptic equation is analytic in the neighborhood of a point where its second derivative exists (LICHTENSTEIN [3.36], see also MÜLLER [Ref.3.37, Lemma 48] and MIRANDA [3.38]). The sum appearing in (3.63) is not only identically zero for values of $\underline{r}$ situated on and outside S, but also identically zero up to $\sigma$:

$$\sum_j b_j \phi_j(\underline{r}) = 0 \quad . \tag{3.64}$$

Then (3.64) can only be valid if $b_j = 0, \forall j$, because the set of functions $\{\phi_n(\underline{r})\}$ is orthonormal. The set of functions $\{\psi_n(\underline{r};k)\}$ is therefore linearly independent, and the numbers $a_n$ may be uniquely determined by for instance orthogonalizing the set of functions $\{\psi_n(\underline{r};k)\}$ or constructing a set of functions $\psi_j^{(1)}(\underline{r};k)$ such that $\{\psi_n, \psi_\ell^{(1)}\}$ is biorthogonal (PELL [3.39]):

$$\int_\sigma \psi_n(\underline{r};k)\psi_j^{(1)}(\underline{r};k)d\sigma = \delta_{n,j} \quad . \tag{3.65}$$

### 3.2.5 The Determination of the Shape of a Scatterer from Far-Field Data

The inverse scattering problem considered here is that of finding the shape of a reflector from the knowledge of the scattered wave. The determination of the shape of a reflector has been accomplished by KELLER [3.40], using the geometrical optical approximation of the field. LEWIS [3.41] developed a wave theory based on the Kirchhoff approximation for scattering, and calculated the shape of the scatterer from the scattered field. PROSSER [3.42] considered the field arising from the scattering of an incoming plane wave by a soft boundary and essentially inverted the Neumann series solution of the integral equation of which the field is a solution. We begin with a review of Keller's geometrical theory.

Suppose a reflector is an arbitrary smooth convex closed surface in the three dimensional space. Let $\sigma(\theta,\phi)$ denote its differential cross section in the direction $\theta,\phi$ due to a plane wave incident from the left along the x-axis. Let $P(\theta,\phi)$ be the point on the reflector from which a ray is reflected in the $\theta,\phi$ direction. Then, according to the laws of reflection, the normal at $P(\theta,\phi)$ must point in the directior $\theta/2,\phi$, and the angle of incidence must be $\theta/2$. The differential cross section $\sigma$ and the reflection coefficient R are related to each other by

$$\sigma(\theta,\phi) = \frac{R(\theta/2)}{4G(\theta/2,\phi)} \quad , \tag{3.66}$$

where G denotes the Gaussian curvature at the point of reflection. Equation (3.66) is the basic relation of the theory, showing that if both R and $\sigma$ are known, the Gaussian curvature of the surface is determined over the hemisphere $0 \leqq \theta \leqq 2^{-1}\pi$ of the unit sphere. The reflectivity is determined from the prior knowledge of the material, and $\sigma$ can be measured. The problem of determining a closed convex surface when its Gaussian curvature is given for every direction of the normal of the surface is known as Minkowski's problem (see MINKOWSKI [3.43] and LEWY [3.44]). An extensive list of references and a detailed analysis concerning the existence of such a surface, provided that

$$\int_{\Omega} G(\underline{n})\underline{n}d\omega = 0 \quad , \tag{3.67}$$

where the integration is carried out over the unit sphere, is found in a paper by NIRENBERG [3.45]. The first uniqueness proof was obtained by Minkowski, and a very simple proof, which requires merely a few derivatives of the surface, has been given by STOKER [3.46].

Unfortunately, (3.66) determines the Gaussian curvature over the hemisphere $0 \leqq \theta \leqq \pi/2$, and therefore any smooth continuation of G over the sphere $0 \leqq \theta \leqq \pi$ subject to (3.67) leads to a reflecting surface. Uniqueness is obtained if the surface is illuminated from the opposite direction, in which way we obtain the values of the Gaussian curvature on the hemisphere $\pi/2 \leqq \theta \leqq \pi$.

As a very simple example of the theory we consider a surface of revolution from which a plane wave incident along the axis of revolution, is scattered (see Fig.3.1). In this case (3.66) with $4G = d\theta/dy$ yields

$$\sigma(\theta) = \frac{R(\theta/2)}{\frac{d\theta}{dy}} \quad . \tag{3.68}$$

Because of

$$\frac{dy}{dx} = \mathrm{cotg}(\theta/2) \quad , \tag{3.69}$$

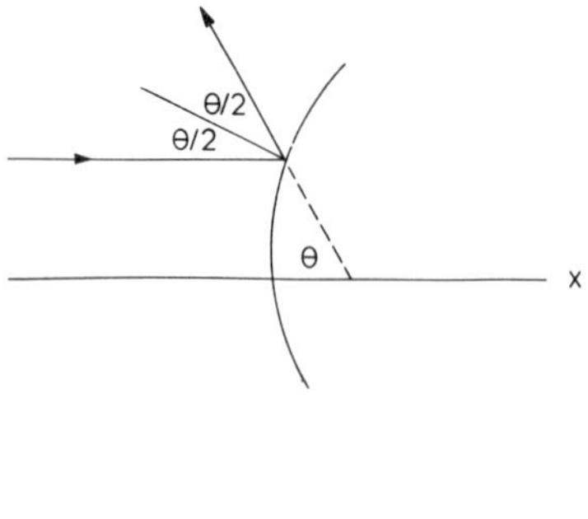


Fig. 3.1. Incident and reflected rays both making the angle $\theta/2$ with the normal to the reflector

we obtain a set of parametric equations for the surface, namely

$$y = y_0 \pm \int_0^\theta \frac{\sigma(\theta')}{R(\theta'/2)}\, d\theta' \quad , \tag{3.70}$$

$$x = x_0 \pm \int_0^\theta \frac{\sigma(\theta')\mathrm{tg}(\theta'/2)}{R(\theta'/2)}\, d\theta' \quad . \tag{3.71}$$

An extensive table with many explicit formulas for special geometries and all kinds of illumination is given by KELLER [3.40]. We will now turn our attention to theories which are not based on the theory of geometrical optics but on the first-order Born approximation (LEWIS [3.41]), also known as the Kirchhoff approximation, of the integral equations describing scattering by a perfect conductor, or on the exact integral equation for the scalar case (PROSSER [3.42]). We will commence with the theory developed by LEWIS [3.41]. Let a plane harmonic electromagnetic wave with electric field vector

$$\underline{E} = \mathrm{Re}\{\underline{E}_0 \exp(i\underline{k}\cdot\underline{r}-i\omega t)\} \tag{3.72}$$

be scattered by a perfectly conducting object with volume $\tau$ and bounded by a closed surface $\sigma$. The backscattered field $\underline{E}_s$ is derived from the representation (3.29); the condition that on the surface of a perfect conductor both $\underline{n}\times\underline{E}$ and $\underline{n}\cdot\underline{H}$ vanishes; the equation

$$\int_\sigma \underline{n}\times\underline{a}\cdot\nabla'\psi d\sigma = \int_\sigma \psi\underline{n}\cdot\nabla'\times\underline{a}\, d\sigma \quad , \tag{3.73}$$

valid for sufficiently smooth functions $\underline{a}$ and $\psi$; and observing that a primed derivative of G is the negative of the unprimed. The result (JONES [Ref.3.31, Sect.8.32]) reads

$$\underline{E}_s(\underline{r}) = (\mathrm{grad\ div}+k^2) \int_\sigma \frac{1}{i\omega}\, \underline{n}\times\underline{H}(\underline{r}')G(\underline{r},\underline{r}';k)d\sigma \quad , \tag{3.74}$$

which for large values of $\underline{r}$ leads to

$$\underline{E}_s(\underline{r}) \simeq \frac{k\exp(ikr)}{4\pi ir} \int_\sigma \{\underline{n}\times\underline{H}(\underline{r}')-[\underline{n}\times\underline{H}(\underline{r}')\cdot\underline{s}]\underline{s}\}\exp(-ik\underline{s}\cdot\underline{r}')d\sigma \quad , \tag{3.75}$$

where $\underline{s}$ denotes the unit vector in the direction of the observation point $\underline{r}$. Using the first-order Born approximation, i.e., taking for $\underline{H}$ the values of the incoming field, (3.75), the relation $\underline{H} = k^{-1}\underline{k}\times\underline{E}$, and observing in the direction $\underline{s} = -\underline{k}$, yields

$$\underline{E}_s(\underline{r}) \simeq \frac{k\exp(ikr)}{4\pi ir}\, \underline{E}_0 \int_{\sigma_1} (\underline{s}\cdot\underline{n})\exp(2i\underline{k}\cdot\underline{r})d\sigma \quad , \tag{3.76}$$

where the surface integral is over the illuminated portion $\sigma_1$ of the surface. From (3.76) and the divergence theorem we obtain

$$\underline{E}_s(\underline{r}) + \underline{E}_s^*(\underline{r}) = \frac{k^2 \exp(ikr)}{2\pi r} \underline{E}_0 \int \gamma(\underline{r}')\exp(2i\underline{k}\cdot\underline{r}')d\underline{r}' \quad , \tag{3.77}$$

with

$$\begin{aligned} \gamma(\underline{r}) &= 1 \quad , \quad \text{if} \quad \underline{r} \in \tau \quad , \\ \gamma(\underline{r}) &= 0 \quad , \quad \text{if} \quad \underline{r} \notin \tau \quad . \end{aligned} \tag{3.78}$$

The integration at the rhs of (3.77) is over the whole three-dimensional space.

If the backscattered field could be measured at all frequencies $\omega = ck$ and for all directions of incidence, then the Fourier transform of $\gamma$ would be known and $\gamma(\underline{r})$ could be determined. The conditions can be mitigated because, due to the finite support of $\gamma$, the lhs of (3.77) is an analytic function of $k_x, k_y, k_z$ (TITCHMARSH [3.47]). The lhs of (3.77) is therefore uniquely determined by its values at any infinite bounded set of points $\underline{k}_j$ (OSGOOD [3.48]). Moreover, the inversion of the Fourier transform of a function with finite support has been considered by HOENDERS and FERWERDA [3.34] who construct an inversion formula only containing the values of the Fourier transform at an arbitrarily chosen bounded infinite set of sampling points.

A formal inversion procedure has been developed by PROSSER [3.42] who improved on the Kirchhoff approximation by considering the full Neumann series solution of the integral equation (3.80). The scattering of a wave function $\psi(\underline{r},k)$ from a soft boundary $\sigma$ is governed by the Helmholtz equation together with the boundary condition

$$\psi(\underline{r},k) = 0 \quad , \quad \text{if} \quad \underline{r} \in \sigma \quad . \tag{3.79}$$

The solution of (3.1) and (3.79) which has to consist of an incoming plane wave plus an outgoing scattered wave is the solution of the integral equation

$$\psi(\underline{r},k) = \exp(i\underline{k}\cdot\underline{x}) + \int_\sigma G(\underline{r},\underline{r}';k) \frac{\partial\psi}{\partial n} (\underline{r}',k)d\sigma' \quad . \tag{3.80}$$

If the Neumann series solution of (3.80) converges, the solution reads

$$\psi(\underline{r},k) = \exp(i\underline{k}\cdot\underline{r}) + \sum_n \int_\sigma G^{(n)}(\underline{r},\underline{r}';k)\exp(i\underline{k}\cdot\underline{r}')d\sigma' \quad , \tag{3.81}$$

where

$$G^{(n)}(\underline{r},\underline{r}';k) = \int_\sigma G^{(n-1)}(\underline{r},\underline{r}'';k)G(\underline{r}'',\underline{r}';k)\frac{\partial}{\partial n(\underline{r}')}d\sigma'' \tag{3.82}$$

and

$$G^{(0)}(\underline{r},\underline{r}';k) = G(\underline{r},\underline{r}';k)\frac{\partial}{\partial n(\underline{r}')} \quad . \tag{3.83}$$

For large values of $|\underline{r}|$ the behavior of $\psi$ is given by

$$\psi(\underline{r},k) \simeq \exp(i\underline{k}\cdot\underline{r}) + (4\pi|\underline{r}|)^{-1}\exp(ikr)T(\underline{k}',\underline{k}) \quad , \tag{3.84}$$

where

$$\underline{k}' = (k/r)\underline{r} \quad , \tag{3.85}$$

and

$$\begin{aligned} T(\underline{k}',\underline{k}) = {} & 2\int_\sigma \exp(-i\underline{k}'\cdot\underline{r}')\frac{\partial}{\partial n(\underline{r}')}\exp(i\underline{k}\cdot\underline{r}')d\sigma' + \dots \\ & + 2^n \int_\sigma \dots \int_\sigma \exp(i\underline{k}'\cdot\underline{r}_1) \cdot \frac{\partial}{\partial n_{(\underline{r}_n)}} \exp(i\underline{k}\cdot\underline{r}_n) \\ & \cdot \prod_{i=1}^{n-1} \frac{\partial}{\partial n(\underline{r}_i)} G(r_i,r_{i+1};k)d\sigma_i \end{aligned} \tag{3.86}$$

Using

$$\int_\sigma hd\sigma = \int_\tau \nabla\gamma(\underline{r}') \cdot \underline{n}\, hd\underline{r}' \tag{3.87}$$

and taking Fourier transforms throughout leads to

$$\begin{aligned} T(\underline{k}',\underline{k}) = {} & 2\gamma(\underline{k}'-\underline{k})(\underline{k}'-\underline{k}) \cdot \underline{k} + \dots + 2^n \int \gamma(\underline{k}'-\underline{k}_2)(\underline{k}'-\underline{k}_2) \cdot \underline{k}_2(\underline{k}_2^2-\underline{k}^2)^{-1} \\ & \cdot \gamma(\underline{k}_2-\underline{k}_3)(\underline{k}_2-\underline{k}_3) \cdot \underline{k}_3(\underline{k}_3^2-\underline{k}^2)^{-1} \dots \gamma(\underline{k}^n-k)(\underline{k}^n-\underline{k}) \cdot kd\underline{k}_n \dots d\underline{k}_1 . \end{aligned} \tag{3.88}$$

Equation (3.88) is a nonlinear integral equation of a type which has been studied for a long time, and is known as an integral power series. Its properties have been discussed by SCHMIDT [3.49], who has shown that (3.88) has a unique solution for sufficiently small values of T, provided that the linear equation $T = 2\gamma\cdot(\ )+\int\gamma\cdot(..)dk$ has a unique solution. Bifurcation of the solution of (3.88) may occur for arbitrary values of T (SCHMIDT [3.49] and VAINBERG and TRENOGIN [3.50]). VAINBERG and TRENOGIN also give a detailed discussion of the modern theory of (3.88). A formal solution of this equation is obtained by PROSSER [3.42] by replacing T by $\varepsilon T$, $\gamma$ by $\varepsilon\gamma$, putting $\underline{k}'$ equal to $-\underline{k}$ and equating equal powers of $\varepsilon^m$. This procedure leads to

$$4k^2\gamma(2\underline{k}) = \sum_{m=1}^{\infty} 4k^2\gamma_m(2k) \tag{3.89}$$

if

$$4k^2\gamma_1(2k) = -T(\underline{k},-k) \quad ,$$

and

$$4k^2\gamma_m(2k) = \sum_{i=2}^{m} \sum_{r_i} 2^i \int \ldots \int \gamma_{r_1}(\underline{k}-\underline{k}_1)(\underline{k}-\underline{k}_1) \cdot \underline{k}_1 \ldots \gamma_{r_i}(\underline{k}_i+\underline{k}) \cdot (\underline{k}_i+\underline{k}) \cdot \underline{k} d\underline{k}_i \ldots d\underline{k}_1 \quad . \tag{3.90}$$

The formal solution uses backscattering data at all energies and all aspects. However, the boundary can be described by two independent parameters, whereas the solution (3.90) uses three independent parameters. It seems therefore plausible that less data will suffice to determine the boundary, which possibility should be further investigated. We finally mention some results obtained by MAJDA [3.51], who has generalized the result of KELLER [3.40] and has shown the possibility of determining the shape of a scatterer with various boundary conditions for the field at its surface from measurements characterized by so-called subsets of determinacy.

## 3.3 Non-Radiating Sources

The problem to be considered in this section concerns the radiation generated by a charge-current distribution. The radiation, radiation reaction, and self-force of a charge-current distribution has been extensively analyzed since the days of SOMMERFELD [3.52], HERGLOTZ [3.53], EHRENFEST [3.54], and SCHOTT [3.11], and it was found that certain distributions do not radiate at all. A famous example of this phenomena is due to SCHOTT [3.11], who has shown that a uniformly charged shell will not radiate while in orbital motion with period T, provided the shell radius is an integral multiple of cT/2, while the orbit need not be circular or even planar. We will present a simple theory constructed by GOEDECKE [3.55] and derive a sufficiency criterion for radiationless distributions in Section 3.3.1. A general theory of nonradiating distributions based on the multipole expansions of an electromagnetic field has been developed by DEVANEY and WOLF [3.13] and is presented in Section 3.3.3. A theory based on integral equations by COHEN and BLEISTEIN [3.14] will be developed in Section 3.3.3. The existence of nonradiating distributions is very important for inverse source problems because it implies that a current-charge distribution cannot be uniquely determined from a measurement of its radiation pattern. In passing, we remark that the radiation pattern suffices to determine the field up to the scatterer, and that therefore the radiation pattern contains all the information of the field outside the scatterer. There is also another reason why nonradiating distributions are important for inverse source problems. Suppose we would consider the scattering of an incoming wave $\underline{E}^{inc}$ by a bounded medium, characterized by a dielectric function $\varepsilon(\underline{r})$. Under the assumption that the Born approximation is valid, the electric field vector reads

$$\underline{E}(\underline{r}) = \underline{E}^{inc}(\underline{r}) + \int_\tau \underline{\underline{G}}(\underline{r},\underline{r}';k)[k^2(n^2-1)\underline{E}^{inc}(\underline{r}')]d\underline{r}' \quad . \tag{3.91}$$

For the meaning of the various symbols we refer to Section 3.3.3. Hence, $k^2(n^2-1)\underline{E}^{inc}$ can be considered as a current distribution leading to the same source term as the integral occurring at the rhs of (3.91), and the conditions for which the integral occurring at the rhs of (3.91) vanishes outside $\tau$ are formulated in Section 3.3.3.

Uniqueness can possibly be obtained using prior information. For instance, COHEN and BLEISTEIN [3.14] have shown that prior knowledge of the time-dependence of a current charge distribution suffices to determine the charge-current distribution from far-field data. (See also SCHMIDT-WEINMAR et al. [3.16].)

### 3.3.1 Early Results and Special Cases

The theory presented in this section has mainly been developed by GOEDECKE [3.55]. Many references are found in the paper by GOEDECKE, as well as in an article by ERBER and PRASTEIN [3.56]. Consider a given charge-current distribution $\underline{j},\rho$, localized in a finite volume V. Suppose that $\underline{j}$ and $\rho$ admit a space-time Fourier decomposition, assuming that the time-dependence of $(\underline{j},\rho)$ is periodic with period T, viz.,

$$\underline{j}(\underline{r},t) = (2\pi)^{-3} \sum_{n=-\infty}^{+\infty} \int_{-\infty}^{+\infty} d\underline{k}\ \exp(-i\underline{k}\cdot\underline{r}+i\omega_n t)\underline{J}(\underline{k},n) \quad , \tag{3.92}$$

$$\rho(\underline{r},t) = (2\pi)^{-3} \sum_{n=-\infty}^{+\infty} \int_{-\infty}^{+\infty} d\underline{k}\ \omega_n^{-1} \exp(-i\underline{k}\cdot\underline{r}+i\omega_n t)\underline{k} \cdot \underline{J}(\underline{k},n) \quad , \tag{3.93}$$

where

$$\omega_n = 2\pi\ nT^{-1} \quad , \quad n = 0,1,2,\ldots \tag{3.94}$$

The Fourier coefficients of $\rho$ follow from the continuity relation. According to classical electromagnetic theory, the retarded potential solutions $(\underline{A},\phi)$ of the Maxwell's equations are (c=1):

$$\underline{A}(\underline{r},t) = \int_\tau d\underline{r}'\ \underline{j}(\underline{r}',t-|\underline{r}-\underline{r}'|)|\underline{r}-\underline{r}'|^{-1} \quad , \tag{3.95}$$

$$\phi(\underline{r},t) = \int_\tau d\underline{r}'\ \rho(\underline{r}',t-|\underline{r}-\underline{r}'|)|\underline{r}-\underline{r}'|^{-1} \tag{3.96}$$

with

$$\underline{E} = -\nabla\phi - \frac{\partial \underline{A}}{\partial t} \quad , \quad \underline{H} = \nabla \times \underline{A} \quad . \tag{3.97}$$

The amount of radiation R is

$$R = \lim_{r\to\infty} \int d\Omega r^2 \hat{\underline{r}} \cdot \underline{S} \quad , \tag{3.98}$$

where $\underline{S}$ denotes the Poynting vector, viz.,

$$\underline{S} = (4\pi)^{-1} \underline{E} \times \underline{H} \quad . \tag{3.99}$$

The leading terms in the asymptotic expansions of (3.95) and (3.96) in a series of inverse powers of $|\underline{r}|$ are:

$$A(\underline{r},t) \simeq |\underline{r}|^{-1} \int_\tau d\underline{r}' \, \underline{j}(\underline{r}',t-|\underline{r}|+\hat{\underline{r}}\cdot\underline{r}') \quad , \tag{3.100}$$

$$\phi(\underline{r},t) \simeq |\underline{r}|^{-1} \int_\tau d\underline{r}' \, \rho(\underline{r}',t-|\underline{r}|+\hat{\underline{r}}\cdot\underline{r}') \quad . \tag{3.101}$$

Inserting (3.100) and (3.101) into (3.97) leads to

$$R = \sum_n \exp i\omega_n(t-|\underline{r}|)\left[\sum_\ell \omega_\ell\omega_{n-\ell} \int_\Omega d\Omega \, \underline{J}(\omega_\ell\hat{\underline{r}},\ell)(\hat{\underline{r}}\hat{\underline{r}}-\underline{\underline{e}})\underline{J}(\omega_{n-\ell}\hat{\underline{r}},n-\ell)\right] \quad . \tag{3.102}$$

A *sufficient* condition for R to be zero for all t is

$$J(\omega_n\hat{\underline{r}},n) = 0 \quad , \quad n = 1,2,\ldots \quad , \tag{3.103}$$

and this simple criterion will be used to construct examples of nonradiating distributions. Suppose that the charge is confined within a sphere with radius b, and that

$$\rho(\underline{r},t) \equiv \rho[\underline{r}-\underline{a}(t)] \quad , \tag{3.104}$$

i.e., the position vector of the center of the sphere is executing a periodic orbit with period T. Because $\underline{j} = \rho\underline{v}$, (3.104) leads to

$$\underline{j} = \dot{\underline{a}}(t)\rho[\underline{r}-\underline{a}(t)] \quad , \tag{3.105}$$

which is consistent with the continuity equation $\nabla\cdot\underline{j} + \frac{\partial\rho}{\partial t} = 0$. Equation (3.103) reduces the problem to finding those $\rho[\underline{r}-\underline{a}(t)]$ for which

$$T^{-1} \int_0^T dt\dot{\underline{a}}(t)\{\exp i[\omega_n\hat{\underline{r}}\cdot\underline{a}(t)-\omega_n t]\} \int_\tau d\underline{r}\rho(\underline{r})\exp(i\omega_n\hat{\underline{r}}\cdot\underline{r}) = 0 \quad , \quad n = 1,2,\ldots \quad , \tag{3.106}$$

or finding those distributions for which the Fourier transform of $\rho$ vanishes at the points $\omega_n \hat{\underline{r}}$, viz.,

$$I(\underline{k}) = \int_\tau d\underline{r}\ \exp(i\underline{k}\cdot\underline{r})\rho(\underline{r}) = 0 \quad , \quad \text{if} \quad \underline{k} = \omega_n \hat{\underline{r}} \quad , \quad n = 1,2,\ldots \quad . \tag{3.107}$$

For a spherically symmetrical charge distribution with radius b, (3.107) reduces to

$$I(\omega_n \hat{\underline{r}}) = 4\pi(\omega_n)^{-1} \int_0^b dr\ r\ \sin(\omega_n r)\rho(r) \quad , \quad n = 1,2,\ldots \quad . \tag{3.108}$$

From (3.108) we obtain Schott's example of a nonradiating spherical shell, which was derived by SCHOTT using an entirely different method, i.e., taking $\rho(y) = \delta(y-b)$ and hence

$$I(\omega_n \hat{\underline{r}}) = 4\pi b\omega_n^{-1}\sin(\omega_n b) \quad , \quad \omega_n = 2\pi n T^{-1} \quad , \quad n = 1,2,\ldots \quad , \tag{3.109}$$

which will be zero if $\omega_1 b = 2\pi b T^{-1} = \ell\pi$, $\ell$ integer $> 0$. This condition is the only requirement, hence the orbits need not be circular nor even planar.

An example of a radiationless volume distribution is

$$\rho(r) = \sum_q A_q \cos(\omega_q r) \quad , \tag{3.110}$$

for $r < b$ and zero outside if q is an integer $\geqq 0$, and the numbers $A_q$ arbitrarily chosen complex numbers. The quantities $I(\omega_n \hat{\underline{r}})$ are zero by the orthogonality property of the circular functions. Examples of spinning and even asymmetric, orbiting, spinning, nonradiating distributions are given by GOEDECKE [3.55].

### 3.3.2 General Theory

The theory of Goedecke was extended by DEVANEY and WOLF [3.13] using the multipole expansions of the electromagnetic field. From the Maxwell equation it follows that the spatial parts $\underline{E}(\underline{r})$ and $\underline{H}(\underline{r})$ of the Cartesian components of the electromagnetic field vectors $\underline{E}(\underline{r},t)$ and $\underline{H}(\underline{r},t)$, generated by a charge-current distribution, with $\underline{E}(\underline{r},t) = \mathrm{Re}\{\underline{E}(\underline{r})\exp(-i\omega t)\}$, and $\underline{H}(\underline{r},t) = \mathrm{Re}\{\underline{H}(\underline{r})\exp(-i\omega t)\}$, satisfy the equations

$$(\nabla^2+k^2)\underline{E}(\underline{r}) = -4\pi[ik\underline{j}(\underline{r})-\nabla\rho(\underline{r})] \quad , \tag{3.111}$$

$$(\nabla^2+k^2)\underline{H}(\underline{r}) = -4\pi(\nabla\times\underline{j}) \quad . \tag{3.112}$$

Therefore, for fields satisfying Sommerfeld's radiation condition at infinity, (3.111) and (3.112) yield

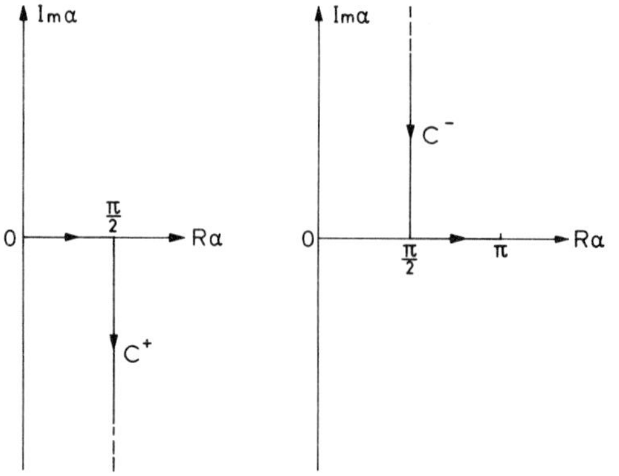


Fig. 3.2. The α-contours of integration $c^+$ and $c^-$

$$\underline{E}(\underline{r}) = -4\pi \int_\tau G(\underline{r},\underline{r}';k)[ik\underline{j}(\underline{r}')-\nabla\rho(\underline{r}')]d\underline{r}' \quad , \tag{3.113}$$

$$\underline{H}(\underline{r}) = -4\pi \int_\tau G(\underline{r},\underline{r}';k)[\nabla\times\underline{j}(\underline{r}')]d\underline{r}' \quad , \tag{3.114}$$

where (WEYL [3.57])

$$G(\underline{r},\underline{r}';k) = \frac{\exp ik|\underline{r}-\underline{r}'|}{4\pi|\underline{r}-\underline{r}'|} = \frac{ik}{8\pi^2}\int_{-\pi}^{+\pi} d\beta \int_{c^\pm} d\alpha \sin\alpha \exp ik\underline{s}\cdot(\underline{r}-\underline{r}') \quad ,$$

and

$$\underline{s} = (\sin\alpha \cos\beta, \sin\alpha \sin\beta, \cos\alpha) \quad . \tag{3.115}$$

The contour $c^+$ applies when $z-z'>0$, and the contour $c^-$ applies if $z-z'<0$ (see Fig.3.2). Combining (3.113,114,115) and using $\nabla\cdot\underline{j}-ik\rho=0$ leads to

$$\underline{E}(\underline{r}) = \frac{ik}{2\pi}\int_{-\pi}^{+\pi} d\beta \int_{c^\pm} d\alpha \sin\alpha\, \hat{\underline{E}}(\underline{s})\exp(ik\underline{s}\cdot\underline{r}) \quad , \tag{3.116}$$

and

$$\underline{H}(\underline{r}) = \frac{ik}{2\pi}\int_{-\pi}^{+\pi} d\beta \int_{c^\pm} d\alpha \sin\alpha\, \hat{\underline{H}}(\underline{s})\exp(ik\underline{s}\cdot\underline{r}) \quad , \tag{3.117}$$

with

$$\hat{\underline{E}}(\underline{s}) = -ik\underline{s} \times [\underline{s}\times\hat{\underline{j}}(k\underline{s})] \quad , \tag{3.118}$$

$$\hat{\underline{H}}(\underline{s}) = ik\underline{s} \times \hat{\underline{j}}(k\underline{s}) \quad , \quad \hat{\underline{j}}(k\underline{s}) = \int_\tau \exp(ik\underline{s}\cdot\underline{r})\underline{j}(\underline{r})d\underline{r} \quad . \tag{3.119}$$

From (3.118) and (3.119) it follows that

$$\underline{s}\cdot\hat{\underline{E}}(\underline{s}) = \underline{s}\cdot\hat{\underline{H}}(\underline{s}) = 0 \quad , \text{and} \quad \hat{\underline{E}}(\underline{s}) = -\underline{s}\times\hat{\underline{H}}(\underline{s}) \quad . \tag{3.120}$$

Consequently, for each $\underline{s}$, the terms in the integrands of (3.116) and (3.117) are plane waves that satisfy the homogeneous Maxwell equation for a monochromatic field with $\omega = ck$ throughout the whole space. Thus (3.116) and (3.117), known as the angular spectrum representation of the electromagnetic field, are throughout their domain of validity a mode expansion of the electromagnetic field. The vector $\hat{\underline{H}}(\underline{s})$ satisfying $\underline{s}\cdot\hat{\underline{H}}(\underline{s}) = 0$ can be expanded into the orthonormal sets of vector spherical harmonics

$$\underline{Y}_\ell^m(\alpha,\beta) = -i\left(\hat{\underline{e}}_\beta\frac{\partial}{\partial\alpha} - \frac{1}{\sin\alpha}\hat{\underline{e}}_\alpha\frac{\partial}{\partial\beta}\right)Y_\ell^m(\alpha,\beta) \tag{3.121}$$

and $\underline{s}\times\underline{Y}_\ell^m(\alpha,\beta)$, which are complete on the unit sphere $\underline{s}\cdot\underline{s} = 1$ for vector fields $\underline{A}$ satisfying $\underline{s}\cdot\underline{A} = 0$ (DEVANEY and WOLF [3.29]). Therefore,

$$\hat{\underline{H}}(\underline{s}) = \sum_{\ell=1}^{\infty}\sum_{m=-\ell}^{+\ell}(-i)^\ell\left[-a_\ell^m\underline{Y}_\ell^m(\alpha,\beta) + b_\ell^m\underline{s}\times\underline{Y}_\ell^m(\alpha,\beta)\right] \quad , \tag{3.122}$$

with

$$a_\ell^m = -\frac{i^\ell}{\ell(\ell+1)}\int_{-\pi}^{+\pi}d\beta\int_0^\pi d\alpha\,\sin\alpha\,\hat{\underline{H}}(\underline{s})\cdot\underline{Y}_\ell^{m*}(\alpha,\beta)$$

and

$$b_\ell^m = -\frac{i^\ell}{\ell(\ell+1)}\int_{-\pi}^{+\pi}d\beta\int_0^\pi d\alpha\,\sin\alpha\,\hat{\underline{H}}(\underline{s})\cdot\left[\underline{s}\times\underline{Y}_\ell^{m*}(\alpha,\beta)\right] \quad . \tag{3.123}$$

The $\ell$ summations begin now with $\ell = 1$ rather than with $\ell = 0$ since there are no vector spherical harmonics of zero degree.

Combining (3.122) with $\hat{\underline{E}}(\underline{s}) = -\hat{\underline{s}}\times\hat{\underline{H}}(\underline{s})$ and using [3.29]

$$h_\ell^{(1)}(kr)Y_\ell^m(\theta,\phi) = \frac{i}{2\pi}(-i)^\ell\int_{-\pi}^{+\pi}d\beta\int_{C^\pm}d\alpha\,\sin\alpha\,Y_\ell^m(\alpha,\beta)\exp(ik\underline{s}\cdot\underline{r}) \tag{3.124}$$

leads to [with (3.17)]

$$\underline{H}(\underline{r}) = \sum_{\ell=1}^{\infty}\sum_{m=-\ell}^{+\ell}\left[-a_\ell^m ik(\nabla\times) + b_\ell^m\nabla\times(\nabla\times)\right]\hat{\underline{r}}h_\ell^{(1)}(kr)Y^m(\theta,\phi)$$

or

$$\underline{H}(\underline{r}) = \sum_{\ell=1}^{\infty}\sum_{m=-\ell}^{+\ell}\left[-a_\ell^m\underline{E}_{\ell,m}^h(\underline{r}) + b_\ell^m\underline{E}_{\ell,m}^e(\underline{r})\right] \quad . \tag{3.125}$$

We therefore succeeded in expanding the field $\underline{H}$ into its vector modes $\underline{E}^h_{\ell,m}(\underline{r})$ and $\underline{E}^e_{\ell,m}$ encountered before [see (3.43) and (3.44)]. The field $\underline{E}$ is calculated from the relation $\nabla \times \underline{H} = \dot{\underline{D}}$, which leads to

$$\underline{E}(\underline{r}) = \sum_{\ell=1}^{\infty} \sum_{m=-\ell}^{+\ell} \left\{a^m_\ell \underline{E}^e_{\ell,m}(\underline{r}) + b^m_\ell \underline{E}^h_{\ell,m}(\underline{r})\right\} \quad . \tag{3.126}$$

It is commonly overlooked that the expansion (3.122) is only valid on the unit sphere $\underline{s} \cdot \underline{s} = 1$, whereas by inserting (3.122) into (3.117) we silently assumed the validity of the expansion (3.122) for *all* values of $\underline{s}$. That the expansion (3.122) is indeed valid for all values of $\underline{s}$ is vital for the theory and is intimately connected with the analyticity properties of the free space field, enabling us to determine the field everywhere from far-field data up to the scatterer. The validity of (3.122) for all values of $\underline{s}$ is readily established, observing that $\hat{\underline{j}}(\underline{s})$ is an analytic function of $\underline{s}$ because it is the Fourier transform of a function with finite support (WHITTAKER and WATSON [3.58]).

The rhs of (3.122) can be shown to be an analytic function of $\underline{s}$, using the asymptotic expansions of the various quantities involved. Therefore, the expansion (3.122), valid for values of $\underline{s}$ satisfying $\underline{s} \cdot \underline{s} = 1$, is by the principle of analytical continuation valid for all complex values of $\underline{s}$.

Using the asymptotic expansion

$$G(\underline{r},\underline{r}';k) \simeq \frac{\exp(ikr)}{4\pi r} \cdot \exp(-ik\underline{r}\cdot\underline{s}') \quad , \tag{3.127}$$

where $\underline{s}'$ denotes a unit vector in the direction of the observation, we obtain from (3.113,114,118,119) the far-field or radiation pattern of the field

$$\underline{E}(\underline{r}) \sim \frac{\exp(ikr)}{r} \hat{\underline{E}}(\underline{s}') \quad , \tag{3.128}$$

$$\underline{H}(\underline{r}) \sim \frac{\exp(ikr)}{r} \underline{s}' \times \hat{\underline{E}}(\underline{s}') \tag{3.129}$$

with

$$\hat{\underline{E}}(\underline{s}') = -ik\underline{s}' \times [\underline{s}' \times \hat{\underline{j}}(k\underline{s}')] \quad . \tag{3.130}$$

It follows from (3.120) and (3.123) that the radiation pattern uniquely determines the coefficients $a^m_\ell$ and $b^m_\ell$, and therefore the field up to charge-current distribution. Moreover, from (3.123) and the relation $\underline{s} \times \hat{\underline{E}}(\underline{s}) = \hat{\underline{H}}(\underline{s})$ we observe that the transverse part $\hat{\underline{j}}^T$ of $\hat{\underline{j}}$, i.e.,

$$\hat{\underline{j}}^T = \frac{(\underline{s}\times\hat{\underline{j}})\times\underline{s}}{|\underline{s}|^2} \quad , \tag{3.131}$$

determines the multipole coefficients $a_\ell^m$ and $b_\ell^m$. For a nonradiating distribution, the integral of the radial component of the time averaged Poynting vector across a sphere with limitingly large radius r vanishes,

$$\langle P\rangle = \frac{1}{8\pi}\,\mathrm{Re}\left\{\lim_{r\to\infty}\int_\Omega d\Omega r^2\hat{\underline{r}}\cdot[\underline{E}(\underline{r})\times\underline{H}^*(\underline{r})]\right\} = 0 \quad . \tag{3.132}$$

Inserting the expansions (3.125) and (3.126) into (3.132) leads to:

$$\langle P\rangle = (8\pi)^{-1}\sum_{\ell=1}^{\infty}\sum_{m=-\ell}^{+\ell}\ell(\ell+1)\left(|a_\ell^m|^2+|b_\ell^m|^2\right) = 0 \quad . \tag{3.133}$$

This can hold only if all the multipole moments $a_\ell^m$ and $b_\ell^m$ vanish. We therefore obtain the result that, according to (3.125) and (3.126) for a nonradiating distribution, the field generated by such a distribution vanishes indentically outside the scatterer. This result has been obtained by ARNETT and GOEDECKE [3.12] for the nonradiating distributions constructed by GOEDECKE [3.55].

We are able to deduce another result from the representations (3.125) and (3.126), namely that the vanishing of $\hat{\underline{j}}^T(k\underline{s})$ for all values of $\underline{s}$ with $\underline{s}\cdot\underline{s}=1$ is a necessary and sufficient condition for a localized charge-current distribution to be nonradiating. The condition is clearly sufficient because (3.119) and (3.123) together with $\underline{s}\times\hat{\underline{E}}=\hat{\underline{H}}$ show that the multipole moments are solely determined by $\hat{\underline{j}}^T$. The necessity of the condition follows from (3.125) and (3.126), and the earlier established result that the field of a nonradiating distribution is identically zero outside the current-charge distribution. A general procedure for obtaining nonradiating distributions, which will also be used for the construction of nonscattering potentials, is formulated by the following theorem [3.13]: if $\underline{f}(\underline{r})$ is any vector field that has continuous partial derivatives up to the third order and that vanishes at all points outside a finite domain D then $\mathrm{Re}\{\hat{\underline{j}}(\underline{r})\exp(-i\omega t)\}$, where

$$\hat{\underline{j}}(\underline{r}) = (4\pi ik)^{-1}[\nabla\times\nabla\times\underline{f}(\underline{r})-k^2\underline{f}(\underline{r})] \tag{3.134}$$

is a localized nonradiating current distribution and $\underline{f}(\underline{r})$ is the space-dependent part of the electric field generated by $\hat{\underline{j}}(\underline{r})$.

The proof of this theorem is obtained from the solution of the vector differential equation

$$\nabla\times\nabla\times\underline{E} - k^2\underline{E} = 4\pi ik\underline{j} \tag{3.135}$$

satisfying the vectorial form of Sommerfeld's radiation condition. This (unique) solution reads

$$\underline{E}(\underline{r}) = \int_{\tau} d\underline{r}' \; 4\pi i k \underline{j}(\underline{r}') \cdot \underline{\underline{G}}(\underline{r},\underline{r}';k) \quad , \tag{3.136}$$

where

$$\underline{\underline{G}}(\underline{r},\underline{r}';k) = (\underline{\underline{e}}+k^{-2}\nabla'\nabla')G(\underline{r},\underline{r}';k) \quad , \tag{3.137}$$

which is a solution of

$$(\nabla\times\nabla\times-k^{2})\underline{\underline{G}}(\underline{r},\underline{r}';k) = \underline{\underline{e}}\delta(\underline{r}-\underline{r}') \quad , \tag{3.138}$$

and $\underline{\underline{e}}$ denotes the unit tensor.

Using (3.135) and

$$\int_{\tau}(\nabla\times\nabla\times\underline{f}\cdot\underline{\underline{G}}-\underline{f}\cdot\nabla\times\nabla\times\underline{\underline{G}})d\tau = \int_{\sigma}[(\underline{n}\times\nabla\times\underline{f})\cdot\underline{\underline{G}}-\underline{f}\cdot(n\times\nabla\times\underline{\underline{G}})]d\sigma \quad , \tag{3.139}$$

which follows from the vectorial form of Green's theorem (3.20) and decomposing the tensor $\underline{\underline{G}}$ into its dyadics yields

$$\underline{E}(\underline{r}) = \underline{f}(\underline{r}) + \int_{\sigma}[(n\times\nabla\times\underline{f})\cdot\underline{\underline{G}}-\underline{f}\cdot(\underline{n}\times\nabla\times\underline{\underline{G}})]d\sigma \quad , \tag{3.140}$$

where $\sigma$ denotes a surface both enclosing $\underline{j}$ and the point $\underline{r}$. However, because $\underline{f}\equiv 0$ if $\underline{r}\notin D$ the rhs of (3.140) vanishes if $\underline{r}\notin D$, showing that $\underline{j}$ is not radiating.

### 3.3.3 Integral Equations and Uniqueness by Prior Knowledge

Another theory of nonradiating sources has been developed by COHEN and BLEISTEIN [3.14], who formulate the problem quite generally using the theory of integral equations. They show, e.g., that prior knowledge about the vectorial properties of the current and its time-dependence in the form

$$\hat{\underline{j}} \cdot \hat{\underline{e}}_z = 0 \quad , \text{and} \quad \underline{j}(\underline{r},t) = \underline{j}(\underline{r})\psi(t) \quad , \tag{3.141}$$

where the spectrum of $\psi(t)$ is not equal to zero in at least a frequency band with finite support, suffices to determine the current distribution uniquely from the values of the electromagnetic field in a surface enclosing $\underline{j}$. The integral equation from which this result is deduced is obtained by substracting from (3.139) a corresponding formula with k replaced by -k, taking $\underline{E}$ for $\underline{f}$, and observing that $\underline{E}$ satisfies (3.135), which yields

$$-i\omega \int_\tau d\underline{r}'\ \underline{j}(\underline{r}',\omega)\cdot(\underline{\underline{e}}+k^{-2}\nabla'\nabla')j_0(k|\underline{r}-\underline{r}'|)$$
$$= \int_\sigma \Big\{\Big(\underline{n}\times\big[\nabla'\times\underline{E}(\underline{r}')\big]\Big)\cdot(\underline{\underline{e}}+k^{-2}\nabla'\nabla')j_0(k|\underline{r}-\underline{r}'|)$$
$$-\underline{E}(\underline{r}')\cdot\Big[\underline{n}\times[\nabla'\times(\underline{\underline{e}}+k^{-2}\nabla'\nabla')]j_0(k|\underline{r}-\underline{r}'|)\Big]\Big\}d\sigma' \quad . \qquad (3.142)$$

Taking the spatial Fourier transform of (3.142) yields

$$-ik(\underline{\underline{e}}-\hat{\underline{k}}\hat{\underline{k}})\hat{\underline{j}}(\underline{k},\omega)\hat{j}_0(\underline{k}) = \hat{j}_0(\underline{k})\Big[\int_\sigma \exp(-i\underline{k}\cdot\underline{r}')\Big\{\underline{n}\times\big[\nabla\times\underline{E}(\underline{r}')\big]\Big\}$$
$$\cdot(\underline{\underline{e}}-\hat{\underline{k}}\hat{\underline{k}})-\underline{E}(\underline{r}')\cdot\Big\{n\times\big[ik\times(\underline{\underline{e}}-\hat{\underline{k}}\hat{\underline{k}})\big]\Big\}d\sigma'\Big] \quad . \qquad (3.143)$$

It appears that we could determine $\hat{\underline{j}}(\underline{k},\omega)$ for all values of $\underline{k}$ cancelling the term $\hat{j}_0(\underline{k})$ appearing at both sides of (3.143). This is not possible because

$$\hat{j}_0(\underline{k}) = \frac{2\pi^2}{\omega^2}\,\delta(k-\omega) \quad , \qquad (3.144)$$

which is zero except on the cone $k=\omega$. However, recalling the assumptions (3.141) for the distribution $\underline{j}(\underline{r},t)$ we observe that $\hat{\underline{j}}^T(\underline{k}) = (\underline{\underline{e}}-\hat{\underline{k}}\hat{\underline{k}})\hat{\underline{j}}(\underline{k})$ can be determined for all values of $\underline{k}$ such that $|\underline{k}|=\omega$, if $\omega$ belongs to the nonzero part of the spectrum of $\psi(t)$, by the rhs of (3.143), which only contains measurable quantities. But $\hat{\underline{j}}^T(\underline{k})$ is the three-dimensional Fourier transform of a function with finite support and is therefore analytic for all complex values of $\underline{k}$. Hence, by the principle of analytical continuation, $\hat{\underline{j}}^T(\underline{k})$ is uniquely determined from its values $\underline{k}$ for which $|\underline{k}|=\omega$ and $\omega$ lying in a nonzero interval. However, because $\hat{\underline{j}}(\underline{k})\cdot\hat{\underline{e}}_z=0$, we can determine $\hat{\underline{j}}(\underline{k})$ uniquely from the values of $\hat{\underline{j}}^T(\underline{k})$, and then calculate $\underline{j}(\underline{r})$ by performing the inverse Fourier transformation.

We next turn our attention to (3.136) and recall that the field of a nonradiating distribution vanishes outside the support of the distribution. Equation (3.136) shows that $\underline{E}(\underline{r})\equiv 0$ for $\underline{r}\notin\tau$ if, integrating by parts,

$$\int_\tau d\underline{r}'\ G(\underline{r},\underline{r}';k)(\underline{\underline{e}}+k^{-2}\nabla'\nabla')\underline{j}(\underline{r}',\omega) = 0 \quad , \quad r\notin\tau \quad . \qquad (3.145)$$

Using the expansion

$$G(\underline{r},\underline{r}';k) = ik\sum_{\ell=0}^{\infty}\sum_{m=-\ell}^{+\ell} h_\ell^{(1)}(kr)j_\ell(kr')Y_\ell^m(\theta,\phi)Y_\ell^{m*}(\theta',\phi') \quad ,$$
$$\text{valid if } r>r' \quad , \qquad (3.146)$$

we derive from (3.145) that, if the current distribution is confined within a sphere with radius a, all the coefficients

$$c_{\ell,m} = \int_{\Omega} d\Omega \int_0^a dr' r'^2 j_\ell(k\underline{r}') Y_\ell^{m*}(\theta',\phi')(\underline{\underline{e}}+k^{-2}\nabla'\nabla')\hat{\underline{j}}(\underline{r}',\omega) \tag{3.147}$$

are zero. The same condition is obtained from (3.142) using the expansion

$$j_0(k|\underline{r}-\underline{r}'|) = ik \sum_{\ell=0}^{\infty} \sum_{m=-\ell}^{+\ell} j_\ell(kr) j_\ell(kr') Y_\ell^m(\theta,\phi) Y_\ell^{m*}(\theta',\phi') \quad . \tag{3.148}$$

Therefore, (3.142) contains the same amount of information on $\underline{j}$ as (3.136) because the same conditions have to be satisfied for those distributions which cannot be observed. To be more specific: the null spaces of the linear transformations $\int_\tau G\, h d\underline{r}$ and $\int_\tau j_0\, h d\underline{r}$, defined as the set of functions h for which these integrals vanish outside $\tau$, are the same for these linear transformations.

Condition (3.147) can be put in a form derived previously, i.e., (3.131), and is also contained in the analysis of GOEDECKE [3.55] via (3.102). From the expansion

$$\exp(-i\underline{k}\cdot\underline{r}) = 4\pi \sum_{\ell=0}^{\infty} \sum_{m=-\ell}^{+\ell} (-i)^\ell j_\ell(kr) Y_\ell^m(\alpha,\beta) Y_\ell^{m*}(\theta,\phi) \quad , \tag{3.149}$$

where $\theta,\phi$ are the polar angles of $\underline{r}$ and $(\alpha,\beta)$ the polar angles of $\underline{k}$, and (3.147) we derive

$$(\underline{\underline{e}}-\hat{\underline{k}}\hat{\underline{k}})\underline{j}(\underline{k},\omega) = 0 \quad \text{if} \quad \omega = k \quad . \tag{3.150}$$

A similar, though simpler, theory has been developed by COHEN and BLEISTEIN [3.14] for the scalar case, and the theory has been generalized for arbitrary time-dependent fields by BLEISTEIN [3.59].

## 3.4 The Determination of an Object from Scattering Data

The main goal of scattering experiments is to obtain information on the object at which an incoming wave is scattered. In this section we will be mainly concerned with the inverse scattering problem connected with the determination of a scalar index of refraction or potential distribution from scattering data. We will assume that the phase of the scattered wave function can be determined from a measurable quantity like the differential cross section or the current density yielding the square of the absolute value of the scattering amplitude. The so-called phase retrieval problem is analyzed in Chapter 2 of this book. For potential scattering, we refer to GERBER and KARPLUS [3.60] and NEWTON [3.61].

Very simple examples already lead to the occurrence of nonuniqueness. For instance, if a plane wave is scattered at a potential well, a discrete set of energies $E_n$ can be found for which no reflection occurs (MESSIAH [Ref.3.62, Chap.3.7]). MORSE and FESHBACH [Ref.3.63, Sect.12.3] obtained a similar result for the potential $V(x) = V_0\{\tanh(xd^{-1})^2\}$, showing that no reflection occurs if

$$(\nu+\frac{1}{4})^{1/2} = n+\frac{1}{2} \quad , \quad n = 0,1,2,\ldots \quad , \tag{3.151}$$

where

$$\nu = 2md^2\hbar^{-2}V_0 \quad . \tag{3.152}$$

It should be observed that the condition (3.151) is not dependent on the energy of the incoming wave.

Similarly if a plane TE electromagnetic wave is incident on a homogeneous dielectric slab, no reflection will occur for certain values of the index of refraction, the thickness of the slab, and the angle of incidence of the wave vector of the incoming plane wave with the normal on the surface (see BORN and WOLF [Ref.3.5, Sect.1.6]). Hence an object cannot be uniquely determined from the values of the scattered field generated by one incoming plane wave and we have either to use prior information on the object or to develop a procedure with which sufficient additional information is obtained. An example of the use of prior information is, for instance, the determination of a spherical symmetrical potential from the knowledge of the energy dependence of a single phase shift. The prior knowledge that the potential only depends on r enables us to determine the potential up to a N-parameter family of equivalent potentials, connected with the so-called bound states (see NEWTON [Ref.3.64, Chap.20]).

Another example of the usefulness of prior information is given by (3.141) and (3.143), showing that prior knowledge of the time-dependence of the current, together with prior knowledge of its vectorial character, ($\underline{j}\cdot\hat{\underline{e}}_z=0$), suffices to determine the current uniquely from its radiation pattern. In this chapter we will discuss two procedures by which a unique determination of the object becomes possible. The first procedure developed by FERWERDA and HOENDERS [3.65] and HOENDERS [3.66] shows that if the validity of the Born approximation is assumed, any potential with finite support can be reconstructed from the knowledge of the scattered fields generated by certain infinite sets of incoming monochromatic plane waves. The second procedure is similar to the method of COHEN and BLEISTEIN (3.141,143), showing that from the knowledge of a set of scattered fields generated by an infinite set of incoming waves with different energies a potential or index of refraction can be reconstructed.

---

$\hbar = h/2\pi$ (normalized Planck's constant)

The different energies (frequencies) and the different directions provide us with an additional degree of freedom. This additional degree of freedom is necessary because the solution of an elliptic equation like the time-dependent Schrödinger equation in the presence of an electrostatic potential is uniquely determined by two parameters, describing the values of the (uniquely determined) solution at a closed surface, whereas the potential depends in general on three parameters. The outline of this chapter is as follows: in Section 3.4.1 we construct several examples of nonscattering dielectrics and potentials and show, for example, the existence of a dielectric slab such that the scattering fields generated by any member of a finite set of incoming plane waves vanishes outside the slab. The existence of an infinite class of dielectrics (or potential distributions) such that every incoming plane wave with fixed frequency and polarization will be transmitted without reflection is shown as well. In Section 3.4.2 we give a brief survey of the inverse scattering problem for spherical symmetrical potentials, and its applications to acoustical and electromagnetic problems will be indicated. In Section 3.4.3 we show that, if the Born approximation is valid, any potential or scalar index of refraction can be reconstructed from the values of the scattered fields generated by an infinite set of incoming monochromatic plane waves with different wave vectors (FERWERDA and HOENDERS [3.65] and HOENDERS [3.66]). Higher-order Born approximations are considered as well, and it is shown that the scattered fields generated by a set of incoming monochromatic plane waves suffices to determine uniquely any potential or index of refraction, (Sect.3.4.4). The reconstruction of an object from scattering data generated by an infinite set of incoming waves with different energies is considered as well.

Section 3.4.5 is devoted to the problem of the analytic continuation of an electromagnetic field to points situated inside the scatterer (WESTON and BOWMAN [3.67], COLTON [3.68], SLEEMAN [3.69]). For example, it is shown that an electromagnetic field can be continued analytically inside perfect conductors with certain geometries like the sphere.

### 3.4.1 Examples of Nonuniqueness

Nonuniqueness is a phenomenon which, unfortunately, arises very often if inverse problems have to be solved. We will construct several examples which are all directly connected with the inversion problems considered in the other sections of this chapter. For instance, *if an incoming wave* $\psi^{\mathrm{inc}}$*, satisfying the three-dimensional Helmholtz equation, is scattered by a potential or scalar index of refraction, an infinite number of potentials or scalar indexes of refraction with finite support can be constructed which do not scatter*. The proof and idea of this theorem is very similar to the characterization of certain classes of nonradiating distributions by DEVANEY and WOLF [3.13] [see (3.134) to (3.140)]. Let a function $\psi^{(c)}$ be equal to $\psi^{\mathrm{inc}}$ if $\underline{r} \notin D$, where D denotes a bounded domain in the three-dimensional x,y,z space. Suppose that $\psi^{(c)}$ be equal to an arbitrarily chosen function not equal to $\psi^{inc}$ if $\underline{r} \in D$, and suppose that $\psi^{(c)}$ is not equal to zero if $\underline{r} \in D$ and twice continuously differentiable with respect to all the variables everywhere in the three-dimensional x,y,z space. Suppose that the potential $V(\underline{r})$ is *defined* by

$$\frac{2m}{\hbar^2} V(\underline{r}) = \frac{(\nabla^2+k^2)\psi^{(c)}}{\psi^{(c)}} \quad . \tag{3.153}$$

The unique solution of the time-independent Schrödinger equation

$$\left(\nabla^2+k^2 - \frac{2m}{\hbar^2} V\right)\psi = 0 \tag{3.154}$$

is the sum of an incoming wave and a scattered wave, satisfying Sommerfeld's radiation condition at infinity (COURANT and HILBERT [Ref.3.70, Vol.II, Chap.4, Sect. 5.2]). However, $\psi^{(c)}$ is a solution of (3.154) satisfying this condition, and therefore, because $\psi^{(c)} = \psi^{inc}$ if $\underline{r} \notin D$ and $\psi = \psi^{(c)}$ the potential (3.153) does not scatter if a wave $\psi^{inc}$ is incident on this potential.

A similar result has been obtained by KAY and MOSES [3.71], who show *that it is possible to construct a physically realizable index of refraction* n(x) *such that plane waves at all angles of incidence but at a fixed frequency and polarization are transmitted without reflection.*

We start with the wave equation in a plane-stratified medium with an index of refraction n(x), $-\infty < x < +\infty$,

$$\left[\nabla^2+k^2n^2(x)\right]\psi(\underline{r}) = 0 \quad . \tag{3.155}$$

For an incident plane wave $\exp(i\underline{k}\cdot\underline{r})$ we can assume that the solution of (3.155) has the form

$$\psi(\underline{r}) = u(x)\exp(ik_y y+ik_z z) \quad , \tag{3.156}$$

and therefore

$$\frac{d^2u}{dx^2} + [E-V(x)]u = 0 \quad , \tag{3.157}$$

where

$$E = k^2-k_x^2-k_y^2 \quad , \text{ and } \quad V(x) = -k^2\left[n^2(x)-1\right] \quad . \tag{3.158}$$

We demand that V(x) must be negative or zero for a physically realizable medium. Equation (3.158) shows that the results obtained for the dielectric slab can be interpreted as well in terms of quantum mechanics: V is a one-dimensional potential which transmits an incident particle with probability one, *no matter what the particle's initial kinetic energy may be*. The problem is to find a function V(x) such that a solution u(x) of (3.157) obeying

$$u(x) \sim \exp(iE^{1/2}x) \quad \text{as} \quad x \to -\infty \quad , \tag{3.159}$$

also fulfills

$$u(x) \sim t(E)\exp(iE^{1/2}x) \quad \text{as} \quad x \to +\infty \quad , \text{ and} \quad |t(E)| = 1 \quad . \tag{3.160}$$

Let us choose N arbitrary positive constants $A_n$ and $\ell_n$ and suppose that a solution of (3.157) has the form

$$u(x) = \left[1 + \sum_{n=1}^{N} \frac{f_n(x)}{\ell_n + iE^{1/2}} \exp(\ell_n x)\right] \exp(iE^{1/2}x) \quad , \tag{3.161}$$

where the functions $f_n(x)$ are the solutions of the N linear equations

$$A_n \exp(\ell_n x) \sum_{\gamma=1}^{N} \exp(\ell_\gamma x) \frac{f_\gamma(x)}{\ell_n + \ell_\gamma} + f_n(x) + A_n \exp(\ell_n x) = 0 \quad , \quad n = 1, \ldots N \quad . \tag{3.162}$$

Substituting (3.161) into (3.157) yields

$$Lu = \left\{ \sum_{n=1}^{N} \left[M_n f_n(x)\right] \frac{\exp(\ell_n x)}{\ell_n + iE^{1/2}} + 2\frac{d}{dx}\left[\sum_{n=1}^{N} f_n(x)\exp(\ell_n x)\right] - V(x) \right.$$
$$\left. \times \exp(iE^{1/2}x) \right\} \quad , \tag{3.163}$$

where

$$L = \frac{d^2}{dx^2} + E - V(x) \quad , \text{ and} \quad M_n = \frac{d^2}{dx^2} - \ell_n^2 - V(x) \quad . \tag{3.164}$$

Equation (3.163) is zero if

$$M_n f_n(x) = 0 \quad , \tag{3.165}$$

and

$$V(x) = 2\frac{d}{dx}\left[\sum_{n=1}^{N} f_n(x)\exp(\ell_n x)\right] \quad . \tag{3.166}$$

We apply the operator $M_n$ to the $n^{th}$ equation of (3.162) and substitute the expression for $V(x)$ given by (3.166). The result is a set of homogeneous linear equations for $M_n f_n(x)$, viz.,

$$M_n f_n(x) + A_n \exp(\ell_n x) \sum_{\gamma=1}^{N} \left[ M_\gamma \frac{f_\gamma(x)}{\ell_\gamma + \ell_n} \right] \exp(\ell_\gamma x) = 0 \quad . \tag{3.167}$$

The coefficient matrix of (3.167) is

$$\| \delta_{n,\gamma} + A_n \frac{\exp(\ell_n + \ell_\gamma)x}{\ell_n + \ell_\gamma} \| = I + A \quad , \tag{3.168}$$

where I denotes the unit matrix, and I + A will be shown to be nonsingular.

On introducing the matrix

$$D = \| A_n^{-1/2} \delta_{n,\gamma} \| \tag{3.169}$$

we obtain

$$\hat{A} = \| A_n^{-1/2} A_\gamma^{-1/2} \exp[(\ell_n + \ell_\gamma)x](\ell_n + \ell_\gamma)^{-1} \| = D\, A\, D^{-1} \quad . \tag{3.170}$$

Hence, $\mathrm{Det}(A+I) = \mathrm{Det}(I+\hat{A})$, and if we have proved that $\det \hat{A} > 0$, the determinant of the matrix (3.168), is not zero, and therefore

$$M_n f_n(x) = 0 \tag{3.171}$$

is the only possible solution of (3.167).
However, if $\hat{A}$ is a positive definite matrix, i.e., given any set of N numbers $y_n$ such that

$$\sum_{n=1}^{N} \sum_{\gamma=1}^{N} y_n y_\gamma^* A_n^{1/2} A_\gamma^{1/2} \frac{\exp(\ell_n + \ell_\gamma)x}{\ell_n + \ell_\gamma} > 0 \quad , \tag{3.172}$$

unless all the $y_n$ are zero, $\mathrm{Det}\, \hat{A} > 0$. But the lhs of (3.172) can be written as

$$\int_{-\infty}^{x} dz \sum_{n=1}^{N} \sum_{\gamma=1}^{N} A_n^{1/2} A_\gamma^{1/2} y_n y_\gamma^* \exp(\ell_n + \ell_\gamma)z = \int_{-\infty}^{x} \left| \sum_{n=1}^{N} A_n^{1/2} y_n \exp(\ell_n z) \right| dz \quad , \tag{3.173}$$

which is always positive unless every $y_n$ is zero, and thus (3.170) to (3.173) show that $\mathrm{Det}(I+A) > 0$, from which the validity of (3.171) follows.

From the solutions of (3.162) for $f_n(x)$ and (3.166) it is readily observed that $V(x)$ approaches zero exponentially if x approaches $+\infty$ or $-\infty$. Moreover, KAY and MOSES [3.71] show that $V(x)$ is negative for all finite x, and (3.159) and (3.160) are immediately deduced from (3.162). *Equation (3.161) enables us to show the existence of an infinite class of potentials vanishing outside a slab* $0 \leqq x \leqq a$, *such that the scattered fields generated by a finite set of, not necessarily monochromatic, plane waves* $\psi_j = \exp(i\underline{k}_j \cdot \underline{r})$, $j=0,1,2,\ldots n$ *incident on the slab, will vanish outside the slab.*

We recall that the field outside the slab is uniquely determined by the values of u and $\partial/\partial x\, u$ at $x = 0$ and $x = a$, and the wave function and its normal derivative are continuous across a boundary. Hence, if at $x = 0$ and $x = a$, u and its derivative are equal to $\psi_j \exp(-ik_y y - ik_z z)$ and $(d/dx)\psi_j \exp(-ik_y y - ik_z z)$, $j=0,1,2,\ldots N$ no scattered field will be observed. Equations (3.161) and (3.162), together with these conditions, leads to a set of equations for the constants $A_n$ and $\ell_n$.

### 3.4.2 Phase Shift Analysis and the Reconstruction of a Potential

If a plane monochromatic wave admitting the expansion (3.140) is scattered by a spherically symmetrical potential the scattered field admits a similar expansion with $j_\ell(kr)$ replaced by a function $u_\ell(r;k)(kr)^{-1}$, where $u_\ell$ is a solution of the radial Schrödinger equation, and

$$u_\ell(r,k) \sim \exp(-i\delta_\ell) \cos\left[kr - \frac{1}{2}\pi(\ell+1) - \delta_\ell\right] \quad , \tag{3.174}$$

where $\delta_\ell$ denotes the so-called phase shift connected with the $\ell$th partial wave. Because $\delta_\ell$ contains information about the potential, the question arises if a potential can be determined from the knowledge of a single-phase shift qua function of the energy. This is indeed possible, up to a N-parameter family of equivalent potentials, connected with the so-called bound states. The problem to which extent a potential can be reconstructed from the knowledge of all the phase shifts for one value of the energy has been treated by NEWTON [Ref.3.64, Sect.20.3].

The procedure to determine a potential from the knowledge of a single phase shift cannot, unfortunately, be generalized to the case of a spherically symmetrical index of refraction (PROSSER [3.42]), because the "potential" $k^2(n^2-1)$ depends on k. A very general procedure with which in principle a bounded, not necessarily spherically symmetrical, potential or index of refraction can be determined, has been proposed by BATES [3.72]. For mathematical convenience we restrict ourselves to two dimensions, and assume that $n^2 - 1 = 0$ if $r \geqq a$. We decompose the wave function $\psi$, which is a solution of (3.155), and $n^2 - 1$ into a Fourier series

$$n^2 - 1 = \sum_{\ell=-\infty}^{+\infty} n_\ell(r)\exp(i\ell\phi) \quad ; \quad \psi = \sum_{\ell=-\infty}^{+\infty} \psi_\ell(r;k)\exp(i\ell\phi) \quad , \tag{3.175}$$

and decompose $\psi_\ell(r)$ into a Dini series (WATSON [Ref.3.73, Chap.18]),

$$\psi_\ell(r;k) = \sum_m b_{\ell,m}(k) J_\ell(h_{\ell,m} r a^{-1}) \quad . \tag{3.176}$$

The numbers $h_{\ell,m}$ are the roots of the equation

$$\psi_\ell(a;k) J_\ell(h_{\ell,m}) = h_{\ell,m} \psi_\ell'(a;k) J_\ell'(h_{\ell,m}) \quad , \tag{3.177}$$

where the prime denotes differentiation with respect to a. These numbers can be determined because $\psi$ can be determined up to the scatterer from far-field data (Sect. 3.2.1). Similarly we can expand $n_\ell(r)$ into a Fourier-Bessel series:

$$n_\ell(r) = \sum_p A_{\ell,p} J_\ell(j_{\ell,p} r a^{-1}) \quad , \text{ if } \quad J_\ell(j_{\ell,p}) = 0 \quad . \tag{3.178}$$

We are therefore led to the determination of the constants $A_{\ell,p}$. Inserting (3.175), (3.176), and (3.178) into (3.155) leads to an infinite system of linear equations for the $b_{\ell,m}$, which possesses only a nontrivial solution if the determinant of the coefficients $b_{\ell,m}$ is equal to zero. The remarkable point is that this determinant only depends on the unknown coefficients $A_{\ell,p}$ and the observed data because the behavior of the wave function within the sphere $r=a$ (characterized by the $b_{\ell,m}$) has been eliminated. It seems reasonable that if the determinant is truncated to order M, and examined for at least M values of k, it is possible to determine m coefficients $A_{\ell,p}$. Further research is needed to decide whether or not this procedure can be used for practical computations.

### 3.4.3 The Determination of a Potential or Index of Refraction from the Scattered Fields Generated by a Set of Monochromatic Plane Waves

If a monochromatic set of plane waves

$$\psi_j = \exp(ik_{j,x}x + ik_y y + im_j z) \quad , \tag{3.179}$$

where

$$m_j = (k^2 - k_{j,x}^2 - k_y^2)^{1/2} \quad , \tag{3.180}$$

and the numbers $k_{j,x}$ are an infinite bounded set of real numbers such that the expression (3.180) is real, is scattered by a potential U with finite support, the scattered wave function $\psi_{sc}$ in first-order Born approximation reads

$$\psi_{sc}(\underline{r}) = \int_\tau d\underline{r}' \; G(\underline{r},\underline{r}';k) V(\underline{r}') \psi_j(\underline{r}') \quad , \tag{3.181}$$

where

$$V(\underline{r}') = \frac{2m}{\hbar^2} U(\underline{r}') \quad . \tag{3.182}$$

Using an expansion due to WEYL [3.57],

$$G(\underline{r},\underline{r}';k) = \frac{ik}{8\pi^2}\iint_{-\infty}^{+\infty} df_x\, df_y (k^2-f_x^2-f_y^2)^{-1/2} \exp\Big[if_x(x-x')$$
$$+if_y(y-y')+i(k^2-f_x^2-f_y^2)^{1/2}|(z-z')|\Big] \quad , \tag{3.183}$$

where $\mathrm{Im}\left\{k^2-f_x^2-f_y^2\right\}^{1/2} \geqq 0$, and taking the Fourier transform of (3.181) with respect to x and y leads to

$$\frac{8\pi^2}{ik}\tilde{\psi}(f_x+k_{j,x},f_y+k_y,z)\left[k^2-(f_x+k_{j,x})^2-(f_y+k_y)^2\right]^{1/2} \exp\left\{-i\left[k^2-(f_x+k_{j,x})^2-(f_y+k_y)^2\right]^{1/2}z\right\}$$
$$= \int_{-a}^{+a} dz' \exp\left\{i\left(k^2-k_{j,x}^2-k_y^2\right)^{1/2}z'-i\left[k^2-(f_x+k_{j,x})^2-(f_y+k_y)^2\right]^{1/2}z'\right\}\tilde{V}(f_x,f_y,z'), \tag{3.184}$$

where the symbol ~ above a function denotes its Fourier transform with respect to x and y and a denotes the support of V along the z axis. It can be shown that the set of functions $\{\exp(i\lambda_j z)\}$ if $\lambda_j=\left(k^2-k_{j,x}^2-k_y^2\right)^{1/2}-\left[k^2-(f_x+k_{j,x})^2-(f_y+k_y)^2\right]^{1/2}$ is complete if $|z|\leqq a$ (PALEY and WIENER [3.74], HOENDERS [3.66], FERWERDA and HOENDERS [3.65]). However, if the set of functions $\{\exp(i\lambda_j z)\}$ is complete if $|z|\leqq a$ then the only function orthogonal to all the functions $\exp(i\lambda_j z)$ vanishes identically (PALEY and WIENER [3.74], FICHTENHOLTZ [3.75]). Hence if there existed another potential $\tilde{V}^{(1)}$ leading to the same lhs of (3.184) for all values of j, $\tilde{V}-\tilde{V}^{(1)}$ would have to vanish identically, showing the desired uniqueness of $\tilde{V}$. The actual determination of $\tilde{V}$ could be achieved by orthonormalizing the set of functions $\{\exp(i\lambda_j z)\}$ if $|z|\leqq a$, and calculating the various Fourier coefficients. HOENDERS [3.66] and FERWERDA and HOENDERS [3.65] obtained a Tannery series for V which only contains the integrals $\int\exp(i\lambda_j z)\tilde{V}(f_x,f_y,z)dz$ as unknowns.

A Tannery series is defined by $\sum_{j=1}^{p} a_j(p)$. Each element $a_j(p)$ depends explicitly on p.

Equation (3.184) has also been derived by WOLF [3.76], who restricted his attention to the homogeneous part of the angular spectrum, i.e., only those values of $f_x$ and $f_y$ were considered such that $k^2-f_x^2-f_y^2>0$. His theory has been extended by DÄNDLIKER and WEISS [3.77], who show that for certain values of $f_x$, $f_y$, and $f_z$ situated within the sphere $f_x^2+f_y^2+f_z^2\leqq k^2$, the three-dimensional Fourier transform $\tilde{V}(f_x,f_y,f_z)$ can be determined from the values of scattered fields in a plane. These scattered fields are, for instance, generated by a set of incoming waves with different energies or different directions of the wave vector. Having shown the possibility of determining a potential or index of refraction if the Born approximation is valid, we now show that a potential or index of refraction with finite support is uniquely determined by the scattered fields generated by the sets of incoming waves mentioned above.

### 3.4.4 The Unique Determination of an Object from Scattering Data

Suppose that an object is not uniquely determined by the scattered fields generated by an infinite set of plane waves (3.179) or plane waves with different energies. Then there would exist a potential or index of refraction such that the scattered fields generated by the set (3.179) vanishes outside the scatterer, and we only observe the incoming field $\psi^{inc}$. The field $\psi$ inside an arbitrary surface $\sigma$ can be represented in terms of its values $\psi^{(s)}$ on the surface $\sigma$:

$$\psi(\underline{r},k) = \int_\sigma \frac{\partial}{\partial_n} K(\underline{r},\underline{r}';k)\psi^{(s)}(\underline{r}')d\underline{r}' \quad , \tag{3.185}$$

where

$$K(\underline{r},\underline{r}';k) = \sum_\ell \frac{\psi_\ell(\underline{r})\psi_\ell(\underline{r}')}{k^2-k_\ell^2} \quad ; \text{ and } \quad (\nabla^2+k_\ell^2+V)\psi_\ell = 0 \quad , \tag{3.186}$$

and the eigenfunctions $\psi_\ell$ are zero at the boundary. We select a surface $\sigma$ such that $k^2$ is equal to one of the eigenvalues $k_\ell^2$.

This is always possible on choosing a surface at a sufficiently large distance d from the origin because, if d tends to infinity, the spectrum becomes continuous. Hence, any small neighborhood around k will, for sufficiently large d, contain an eigenvalue k, which by a suitable boundary perturbation (MORSE and FESHBACH [Ref.3.63, Vol.II, Sect.9.2], WASSERMAN [3.78,79]) can be made equal to $k^2$.

If $k^2 = k_\ell^2$, the theory of partial differential equations requires that

$$I \equiv \int_\sigma \frac{\partial}{\partial_n} \psi_\ell(\underline{r}')\psi_j(\underline{r}')d\sigma = \int_\tau V(\underline{r}')\psi_\ell(\underline{r}')\psi_j(\underline{r}')d\underline{r}' = 0 \quad , \tag{3.187}$$

as suggested by (3.185) and (3.186) and taking $\psi^{(s)} = \psi_j$.

The volume integral is immediately derived from the surface integral using Green's theorem and $\tau$ denotes the support of V. The function $\psi = \exp\left[iax+if_y+i(k^2-f_y^2-a^2)^{1/2}z\right]$, where a is situated within an arbitrary bounded domain D, not containing the branch points $a = \pm(k^2+f_y^2)^{1/2}$, can be approximated arbitrarily closely and uniformly for all values of $x,y,z \in \tau$ by a linear combination of a sufficient large number of functions $\psi_j$ and Lagrangian interpolation polynomials.

The theory of Lagrangian interpolation (FERWERDA and HOENDERS [3.34,3.80], MARKUSHEVICH [Ref.3.81, Vol.II, probl.(2.30)]) tells us that this statement is true if a is situated within a sufficiently small circular domain C. Hence, if D is covered with a (*finite*) number of domains C, which is possible because the radius of C only depends on its distance to a, repeated application of Lagrangian interpolation leads to the desired result.

Choosing $\mathrm{Im}\{k^2-a^2-f_y^2)^{1/2}\} \geqq 0$, a real, the main contribution to the integral (3.187) if $a\to\infty$ arises from small domains A around those parts of the boundary surface S of $\tau$ where z is maximal ($z=z_m$ say). The potential $V(\underline{r})$ is by definition not identically zero if $\underline{r}\in A\cap\tau$, whereas $\psi_\ell(\underline{r})$ is also not identically zero if $\underline{r}\in A\cap\tau$ because this would imply that $\psi_\ell(\underline{r})$, which is a regular solution of an elliptic equation, would vanish everywhere inside $\sigma$. If we restrict ourselves to surfaces for which $z=z_m$ for a finite number of points $(x_j,y_j)$, $j=1,\ldots p$, $\in S$, the asymptotic expansion of I reads

$$I \sim \sum_{j=1}^{p} A_j\, a^{-\alpha}\exp\left[iax_j+ik_y y_j+i(k^2-a^2-k_y^2)^{1/2} z_m\right] \quad, \text{ if } \quad a\to\infty \quad, \tag{3.188}$$

where the complex numbers $A_j$ are not equal to zero if $j=1,\ldots p$ and $\alpha$ denotes a positive number. But the rhs of (3.188) is not identically zero if a belongs to any finite interval $\ell$ of the real a-axis because the set of functions $\exp(iax_j)$ is linearly independent if $a\in\ell$. Therefore, the integral (3.187) can only vanish for all incoming waves $\psi_\ell$ if V is identically zero. If the value $z_m$ is attained for a continuum of points x and y, a much more difficult analysis leads to the same result.

Hence, a potential is uniquely determined by the scattered fields outside $\tau$, generated by the waves (3.179), because if two different potentials led to the same scattered fields, their difference, leading to vanishing scattered fields, would be zero. In exactly the same way we are led to the conclusion that a potential with finite support is uniquely determined by the scattered fields generated by a set of incoming plane waves with different energies. This result is obtained considering the solution $\psi(\underline{r};k)=\psi^{inc}(\underline{r};k)+\int_\tau\Gamma(\underline{r},\underline{r}';k)\psi^{inc}(\underline{r}';k)d\underline{r}'$ of the scattering integral equation $\psi=\psi^{inc}+\int_\tau GV\psi d\underline{r}'$, where $|\Gamma(\underline{r},\underline{r}';k)=O(k^2)$ if $|k|\to\infty$, $0<\arg k<2\pi$, and both $\underline{r}$ and $\underline{r}'\in\tau$ (HOENDERS [3.82]). The Green's function $\Gamma$ admits also an expansion into the natural modes of the scattering problem.

*The results of this section are also valid if, instead of by a potential, an object is characterized by an index of refraction* because the Green's function $\Gamma$ admits the same expansion as formulated by (3.186) (COURANT and HILBERT [Ref.3.70, Vol.I, Chap.5, Sect.14.3]).

### 3.4.5 The Analytical Continuation of the Electromagnetic Field from the Exterior to the Interior of a Scatterer and Its Physical Implications

It was shown that the reconstruction of an index of refraction from the scattered fields generated by a finite set of incoming plane waves is in general not unique. However, a unique reconstruction of the object might become possible given a sufficient amount of prior knowledge about the scatterer. For instance, the prior knowledge that the field can be continued analytically for values of its argument situated inside the scatterer might restrict the class of solutions of the reconstruction problem. However, it will be shown that for the case of a perfectly conducting body the analytical continuation of the electromagnetic field is possible for a large number of certain geometries for a nonempty set of points situated inside the scatterer, whereas the field inside the conductor is identically zero (WESTON et al. [3.67]). This example shows that the field obtained by analytical continuation differs from the physical field. Moreover, from Weierstrass's theorem of the permanence of a functional equation (HILLE [Ref.3.83, Vol.II, Sect.10.7], OSGOOD [Ref.3.84, Chap.9, Sect.6]), we deduce that the values of the field obtained by analytical continuation still satisfy the free space Maxwell equations and therefore always differ from the existing fields. We will formulate Weston's main theorem as follows:

*Let an arbitrary plane* $z = z_0$ *intersect a perfectly conducting scattering surface* S *and denote the portion of* S *above this plane by* $S_0$*. If* $S_0$ *is convex there exists a domain* D *inside the scatterer, which, depending on* S*, can be empty, such that the vector potential* $\underline{A}$ *can be continued analytically for values of its argument* $\in$ D.

The proof of this theorem obtained from the representation theorem (3.29), the condition $\underline{n} \times \underline{E} = 0$ valid on the surface of a perfect conductor, the procedure given by (3.73) observing that the term grad div(...) can be omitted because $\underline{E}$ and $\underline{H}$ are related to $\underline{A}$ by $\underline{E} = ik^{-1}(\nabla\times\nabla\times\underline{A})$ and $\underline{H} = \nabla\times\underline{A}$ and the expansion (3.115) which leads to

$$\underline{A}(\underline{r}) = \frac{ik}{8\pi^2}\int_{-\pi}^{+\pi} d\beta \int_{C^+} d\alpha\, \sin\alpha\, \exp(ik\underline{s}\cdot\underline{r})\left[\int_S \underline{n}\times\underline{H}(\underline{r}')\exp(-ik\underline{s}\cdot\underline{r}')d\sigma'\right] \quad . \tag{3.189}$$

The analyticity properties of the expression (3.189) are determined by the integral over S. We set $\alpha = 2^{-1}\pi - it$ and choose the coordinate system such that the origin is on S with the positive z-axis pointing outwards, and orient the system so as to have $s\cdot x = x\cosh t + z\sinh t$. The dominant contribution of the integral over S comes from the neighborhood of the point $y = y_0$, where $\partial f/\partial y_0 = 0$, and by means of saddle point integration we have

$$\int_S \underline{n}\times\underline{H}(\underline{r})\exp(-ik\underline{s}\cdot\underline{r})d\sigma \sim \left(\frac{2\pi}{k\sinh t}\right)^{1/2}\int_{x_1}^{x_2} \exp\left\{-ik\left[x\cosh t + if(x,y_0)\sinh t\right]\right\} \cdot \underline{n}\times\underline{H}(\underline{r})\Big|_{y=y_0} dx \tag{3.190}$$

where $z = f(x,y)$ describes S.
The asymptotic evaluation of (3.190) for large values of t is obtained from a theorem of VAN DER CORPUT [3.85] (see ERDÉLYI [Ref.3.86, Sect.2.3]) which states that the main contribution to the integral arises from those points p at which $(-ik\sinh t)^{1/2}h'(x)[h''(x)]^{-1/2}$ is real, while the imaginary part of this function changes its sign when x passes through p, where $h = \mathrm{cotgh}\, t + if(x,y_0)$. Hence,

$$\int_S \underline{n}\times\underline{H}(\underline{r}')\exp(-ik\underline{s}\cdot\underline{r}')d\sigma = O[\exp(kp\sinh t)] \quad , \text{if} \quad t\to\infty \quad , \tag{3.191}$$

and the representation (3.189) converges and is analytic if $z > p$.

As an application we consider a surface of revolution $z = -b + b[1 - a^{-2}(x^2+y^2)]^{1/2}$ and take the z-axis to be the axis of revolution. Using the proof of the theorem we deduce that $\underline{A}(\underline{r})$ can be continued analytically for all values of $\underline{r}$ inside the surface. In this regard, we recall that the field representation given by the Mie series solution is analytic everywhere in space (WESTON [3.67]). Similar theorems have been derived by COLTON [3.68] and SLEEMAN [3.69].

## References


3.1 R. Hosemann, S.N. Bagchi: *Direct Analysis of Matter by Diffraction* (North-Holland, Amsterdam 1963)
3.2 R.E. Langer: Bull. Amer. Math. Soc. (2) *39*, 814-820 (1933)
3.3 M. Kac: Am. Math. Monthly *73*, 1 (1966)
3.4 H.P. Baltes, E.R. Hilf: *Spectra of Finite Systems* (Bibliographisches Institut, Zürich 1976)
3.5 M. Born, E. Wolf: *Principles of Optics*, 3rd ed. (Pergamon Press, Oxford 1965)
3.6 P. Debeye: Ann. d. Physik (4) *30*, 73 (1909)
3.7 G. Borg: Acta Math. *78*, 1-94 (1946)
3.8 V. Ambarzumian: Z. Physik *53*, 690-695 (1929)
3.9 R.G. Newton: Siam Rev. *12*, 346-356 (1970) (review)
3.10 H.P. Baltes: Appl. Phys. *12*, 221-244 (1977), Sect.3.1 (review)
3.11 G.A. Schott: Phil. Mag. (Suppl.) *15*, 752-761 (1933)
3.12 J.B. Arnett, G.H. Goedecke: Phys. Rev. *168*, 1424-1428 (1968)
3.13 A.J. Devaney, E. Wolf: Phys. Rev. D *8*, 1044-1047 (1973)
3.14 J.K. Cohen, N. Bleistein: J. Math. Phys. *18*, 194-201 (1977)
3.15 R. Mireless: J. Math. and Phys. *45*, 179-187 (1966)
3.16 H.G. Schmidt-Weinmar, D.K. Lam, A. Wouk: Can. J. Phys. *54*, 1925-1936 (1976)
3.17 A.M. Cormack: J. Appl. Phys. *34*, 2722-2727 (1963)
3.18 J. Radon: Ber. d. Math. Phys. Kl. d. Sächs. Ges. d. Wiss. Leipzig *69*, 262-277 (1917)
3.19 Ph. Mader: Math. Z. *26*, 646-652 (1927)
3.20 G.E. Uhlenbeck: Physica *5*, 423-428 (1925)
3.21 J.M. Gel'fand, G.E. Schilow: *Verallgemeinerte Funktionen* (Deutscher Verlag der Wissenschaften, Berlin 1960)
3.22 M. Zwick, E. Zeitler: Optik *38*, 550-565 (1973)
3.23 R.B. Marr: J. Math. Anal. Appl. *45*, 357-374 (1974)
3.24 E. Zeitler: Optik *39*, 396-415 (1974)
3.25 V.F. Turchin, V.P. Kozlov, M.S. Malkevich: Sov. Phys.-Usp. *13*, 681-703 (1971)
3.26 M.M. Lavrentiev: *Some Improperly Posed Problems of Mathematical Physics* (Springer, Berlin, Heidelberg, New York 1967)
3.27 A. Sommerfeld: *Partial Differential Equations in Physics*, Lectures on Theoretical Physics, Vol.VI, 5th printing (Academic Press, New York 1967) Sect.28.6, pp. 191-192
3.28 C.H. Wilcox: Comm. on Pure and Appl. Math. *9*, 115-134 (1956)
3.29 A.J. Devaney, E. Wolf: J. Math. Phys. *15*, 234-244 (1974)
3.30 J.M. Blatt, V.F. Weisskopf: *Theoretical Nuclear Physics* (Wiley and Sons, New York 1966) pp. 798-799
3.31 D.S. Jones: *The Theory of Electromagnetism* (Pergamon Press, Oxford 1964)
3.32 J.R. Shewell, E. Wolf: J. Opt. Soc. Am. *58*, 1596-1603 (1968)
3.33 G.C. Sherman: J. Opt. Am. *57*, 1490-1498 (1967)
3.34 B.J. Hoenders, H.A. Ferwerda: Optik *37*, 542-556 (1973)
3.35 K. Miyamoto: J. Opt. Soc. Am. *50*, 856-858 (1960)
3.36 L. Lichtenstein: *Encyclopädie der Mathematischen Wissenschaften*, Vol.II-3-2 (Teubner, Leipzig 1923) Chap.12
3.37 C. Müller: *Grundprobleme der Mathematischen Theorie Elektromagnetischer Schwingungen*. Grundlehren der Mathematischen Wissenschaften, Vol.88 (Springer, Berlin, Göttingen, Heidelberg 1957) Lemma 48

3.38 C. Miranda: *Equazioni alle Derivate Parziali di Tipo Ellittico*. Ergebnisse der Mathematik und ihrer Grenzgebiete (Springer, Berlin, Göttingen, Heidelberg 1955) Sect.44
3.39 A. Pell: Trans. Am. Math. Soc. (5) *12*, 135-164 (1911)
3.40 J.B. Keller: IRE Trans. AP, 146-149 (1959)
3.41 R.M. Lewis: IEEE Trans. AP-*17*, 308-314 (1969)
3.42 R.T. Prosser: J. Math. Phys. *10*, 1819-1822 (1969)
3.43 H. Minkowski: Math. Ann. *57*, 447-495 (1903)
3.44 H. Lewy: Trans. Am. Math. Soc. *43*, 258-270 (1938)
3.45 L. Nirenberg: Comm. on Pure and Appl. Math. *6*, 337-394 (1953)
3.46 J.J. Stoker: Comm. on Pure and Appl. Math. *3*, 231 (1950)
3.47 E.C. Titchmarsh: *The Theory of Functions*, 2nd ed. (Oxford University Press, 1939) Sect.2.83
3.48 W.F. Osgood: *Lehrbuch der Funktionentheorie*, Vol.2 (Teubner, Leipzig 1929) Sect.12
3.49 E. Schmidt: Math. Ann. *65*, 370-399 (1908)
3.50 M.M. Vainberg, V.A. Trenogin: *Theory of Branching of Solutions of Non-Linear Equations* (Noordhoff, Leiden 1974)
3.51 A. Majda: Comm. on Pure and Appl. Math. *29*, 261-291 (1976)
3.52 A. Sommerfeld: Nachr. Akad. Wiss. Göttingen, Math. Phys. Kl.IIa, Math. Phys. Chem. Abt. p.99, 363 (1904); p.201 (1905)
3.53 G. Herglotz: Math. Ann. *65*, 87 (1908)
3.54 P. Ehrenfest: Phys. Z. *11*, 708 (1910)
3.55 G.H. Goedecke: Phys. Rev. *135*, B281-288 (1964)
3.56 T. Erber, S.M. Prastein: Act. Phys. Austr. *32*, 224-243 (1970)
3.57 H. Weyl: Ann. Phys. (Paris) *60*, 481 (1919)
3.58 E.T. Whittaker, G.N. Watson: *A Course of Modern Analysis*, 4th ed. (Cambridge University Press, 1963) Sect.5.31
3.59 N. Bleistein: J. Acoust. Soc. Am. *60*, 1249-1255 (1976)
3.60 R.B. Gerber, M. Karplus: Phys. Rev. D *1*, 998-1012 (1970)
3.61 R.G. Newton: J. Math. Phys. *9*, 2050 (1968)
3.62 A. Messiah: *Quantum Mechanics*, Vol.1 (North-Holland, Amsterdam 1965) Chap.3, Sect.7
3.63 Ph.M. Morse, H. Feshbach: *Methods of Theoretical Physics* (McGraw-Hill, New York 1953)
3.64 R.G. Newton: *Scattering Theory of Waves and Particles* (McGraw-Hill, New York 1966)
3.65 H.A. Ferwerda, B.J. Hoenders: Optik *39*, 317-326 (1974)
3.66 B.J. Hoenders: Ph. D. Thesis, State University at Groningen (1972) Chap.7
3.67 V.H. Weston, J.J. Bowman, Ergun Ar: Arch. Rat. Mech. Anal. *31*, 199-213 (1968)
3.68 D. Colton: Quart. J. Math. Oxford (2) *22*, 125-130 (1971)
3.69 B.D. Sleeman: Proc. Camb. Phil. Soc. *73*, 477-488 (1973)
3.70 R. Courant, D. Hilbert: *Methods of Mathematical Physics*, 3rd printing (Interscience, New York 1966)
3.71 I. Kay, H.E. Moses: J. Appl. Phys. *27*, 1503-1508 (1956)
3.72 R.H.T. Bates: J. Phys. A *8*, L80-82 (1975)
3.73 G.N. Watson: *A Treatise on the Theory of Besselfunctions* (Cambridge University Press 1966)
3.74 R.E.A.C. Paley, N. Wiener: *Fourier Transforms in the Complex Domain*. American Mathematical Society Colloquium Publications, Vol.19 (Am. Math. Soc. Providence R.I. 1934) Sect.11
3.75 G. Fichtenholtz: Rend. Circ. Mat. Pal. *50*, 385-398 (1926)
3.76 E. Wolf: Opt. Comm. *1*, 153-156 (1969)
3.77 R. Dändliker, K. Weiss: Opt. Commun. *1*, 323-328 (1970)
3.78 G.D. Wasserman: Phil. Mag. *37*, 563-570 (1946)
3.79 G.D. Wasserman: Proc. Camb. Phil. Soc. *44*, 251-262 (1948)
3.80 H.A. Ferwerda, B.J. Hoenders: Optik *40*, 14-17 (1974)
3.81 M. Markushevich: *Theory of Functions of a Complex Variable*, Vol.2 (Prentice-Hall, New York 1965)
3.82 B.J. Hoenders: "On the Decomposition of the Electromagnetic Field into Its Natural Modes, in *Proc. 4th Rochester Conf. on Coherence and Quantum Optics*, June 8-10, 1977, ed. by L. Mandel, E. Wolf (Plenum Press, New York 1978)

3.83 E. Hille: *Analytic Function Theory*, Vol.2 (Chelsea, New York 1973)
3.84 W.F. Osgood: *Lehrbuch der Funktionentheorie*, Vol.1 (Teubner, Leipzig 1920)
3.85 J.G. Van der Corput: Comp. Math. *3*, 328-372 (1936)
3.86 A. Erdélyi: *Asymptotic Expansions* (Dover, New York 1956)
3.87 B.J. Hoenders: Completeness of a set of modes connected with the electromagnetic field of a homogeneous sphere embedded in an infinite medium. To be published in J. Phys. A
3.88 A.J. Devaney: J. Math. Phys. *19*, 1526-1531 (1978)

# 4. Spatial Resolution of Subwavelength Sources from Optical Far-Zone Data

H. G. Schmidt-Weinmar

With 17 Figures

Superresolution, i.e., the reconstruction of source details of the order of a wavelength and below, is an old and controversial question. Much of the renewed interest in this problem is motivated by potential applications in, for example, medical and microbiological research on the one hand, and by the availability of powerful tools (lasers, data acquisition systems, fast computers) on the other hand.

The three main aspects of optical reconstruction problems described in Chapter 3, namely the determination of the phase, the reconstruction of the field, and the reconstruction of the scattering object, are likewise important here. The difficulties involved in "superresolving" reconstruction are still more pronounced than those of "general" reconstruction problems described in Chapters 2 and 3. Promising studies of superresolution problems therefore appear to be feasible only in well-specified model situations, where a sufficient amount of prior knowledge and constraints can be taken for granted.

Since comprehensive articles on earlier approaches are available (see Sect.4.1), in the present chapter we emphasize one particular superresolution problem of current interest. We review the recent progress in the *reconstruction of the near field* of sources which are, by prior knowledge, *bounded in space and located in a region with linear dimensions of the order of a wavelength*. We also discuss a possible solution of the pertinent phase problem. Moreover, we describe the representation of wave fields by nonuniform plane waves. This representation is basic to our approach.

Various old and new studies of superresolution are briefly reviewed in Section 4.1. Waves associated with complex-valued spatial frequencies and the pertinent representation of fields from localized sources are described in Sections 4.2 and 4.3. In Sections 4.4 and 4.5 we discuss the errors that result from any band-limiting of the field of a subwavelength source. Next, in Section 4.6 we describe how the spectrum of complex-valued spatial frequencies of the near field of a subwavelength source is specified from the far-zone data. Finally, in Section 4.7 we discuss a recent proposal for the measurement of the far field, including the determination of the spatial distribution of the phase.

## 4.1 Approaches to Superresolution

Superresolution from optical far-zone information has been considered practically impossible, chiefly for the following three reasons.

1) *Measurement.* If the spatial distribution of a wave field at large radial distances from a scattering or diffracting source does indeed contain significant information about the spatial distribution of the near field, the measurement will require high precision; for example, a length will have to be measured with a tolerance $\ll \lambda$, where $\lambda$ is the wavelength of light. Since optical quantities have to be measured through the energy of radiation rather than through the magnitude and the phase of the electromagnetic field, a method that involves the measurement of the spatial distribution of the phase of a scattered or diffracted field (see Chap.2) is likely to be sophisticated. Although these are strong arguments against practicable superresolution from optical far-zone data, there is an interferometric method to measure the data as required which is discussed in Section 4.7.

2) *Stability.* The problem of the near-field reconstruction from far-zone data is known to be improperly posed in the sense that the solutions do not depend continuously on data. Stability can, however, be restored by introducing suitable stabilizing constraints upon the field.

3) *Information.* In the far zone of a localized source, any wave field is approximated at any field point by a homogeneous (or uniform) plane wave (see Fig.4.1); apparently there is no feasible measurement in the far zone that can acquire information other than what is associated with the homogeneous plane-wave components of a given field. Clearly, this view tends to question the possibility of superresolution from far-zone data and leads one to conclude that, since no more information is available in the far zone than what is contained in a spectrum of homogeneous plane waves, to obtain superresolution one has to measure the spatial distribution of the field in the near zone or continue analytically the spectrum of the homogeneous plane waves.

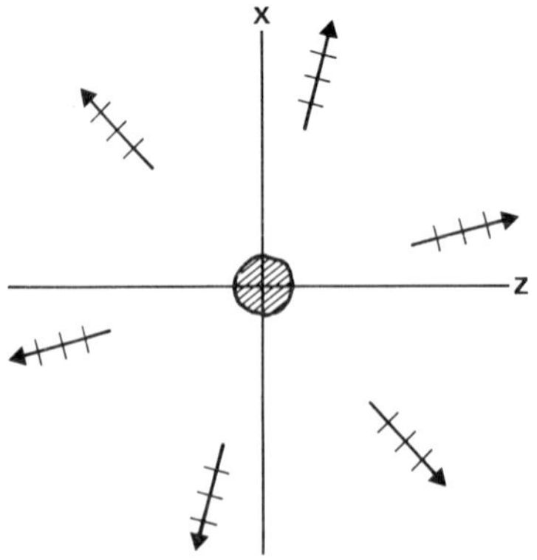


Fig. 4.1. Approximation by homogeneous plane waves of the field at a far-zone point

The following is an intuitive argument in support of the possibility of superresolution from far-zone data. The object far field is close to a spherical wave, if it is transmitted from a subwavelength source; if one can measure the spatial distribution of the field of a spherical wave, it is possible (MILLAR [4.1]) to deduce, from this distribution, the position of the singularity of this wave field. This view has led us to surmise that the spatial distribution of subwavelength sources can be resolved from far-zone data to within a small fraction of a wavelength.

It is well known [4.2] that the Fourier transform of a field that has finite support in one of the planes $Z^+$ and $Z^-$ (see Fig.4.2) is an entire analytic function of the planar spatial frequencies $f_x$, $f_y$. Superresolution has thus frequently been considered the problem of how to extrapolate or continue analytically a spectrum of planar spatial frequencies from its known values inside a given low-frequency domain into a high-frequency region. For references to this problem the reader is referred to review articles by GOODMAN [4.3], HUANG et al. [4.4], FRIEDEN [4.5], and a recent paper by PASK [4.6]. LUKOSZ's approach is described in, e.g., [4.3] (see also FERWERDA [4.7]). VIANO [4.8] noted that object reconstruction from finite-aperture data is an improperly posed problem, and has given a method of restoring stability by introducing additional constraints. WOLTER [4.9,10,11] and HOENDERS and FERWERDA [4.12] studied the sampling theorem with regard to the number of degrees of freedom of an image. All those authors agree on the theoretical possibility of superresolution from far-zone data for bounded objects and noiseless imaging systems; but if the measured low-frequency spectrum is noisy, it is practically impossible to find, from the low-frequency data, the spectrum outside a given low-frequency domain.

SLEPIAN and POLLAK [4.2], LANDAU and POLLAK [4.13,14], and SLEPIAN [4.15] decomposed the spectrum in terms of the doubly-orthogonal prolate spheroidal wave functions and concluded that analytic continuation of a spectrum into a high-frequency region is unfeasible when the low-frequency part of the spectrum is measured in the presence of noise. This result agrees with the uncertainty relation; if the nonzero amplitude of a planar source is confined essentially to within an area $\Delta x \Delta y$ that is small compared to the squared wavelength, the magnitude of its planar-spatial-frequency spectrum is very small inside the low-frequency domain $\Delta f_x \Delta f_y \leqq 1/\lambda^2$. Thus, although the spectrum of planar spatial frequencies of the field of a source with finite support is given by one entire analytic function, insufficient information on high spatial frequencies is found inside a low-frequency domain.

### 4.1.1 Array of Sources with Known Radiation Pattern

WOLTER and co-workers [4.16] demonstrated the possibility of superresolution from far-zone data, if an rf EM field arises from an array of a finite number of dipoles whose radiation patterns are known a priori. Wolter's resolution from far-zone data of the array of dipoles was not diffraction-limited, and high precision was not required of the measurements. The resolution of the array of dipoles was obtained directly from the spatial distribution of the far field without any representation of the field by an angular plane-wave spectrum. PULVERMACHER [4.17,18] demonstrated superresolution from optical data for periodic objects.

Wolter's experiment has shown in particular that there is a possibility of practical superresolution from far-field data, if it is known a priori that the source is composed of point-source dipoles; more information than is associated with homogeneous plane waves must then be available in the far zone. Wolter's results lead us to the following questions.

I) Under which prior knowledge is superresolution feasible with *continuous* sources, in particular with continuous *planar* sources?

II) If the radiation pattern of the elementary sources of an array is completely or partly *unknown*, what spatial resolution power can be obtained from far-zone data and what precision is required of the measurement?

III) How is an *optical* system to be designed that is not diffraction-limited?

### 4.1.2 Superresolution Using Evanescent Waves

NASSENSTEIN [4.19,20] achieved superresolution using the diffraction of evanescent plane waves to translate spatial frequencies in excess of $1/\lambda$ into the conventional optical region. NASSENSTEIN [4.20-22], LUKOSZ and WÜTHRICH [4.23] also used evanescent plane waves with holography.

### 4.1.3 $\lambda$-Localized Sources

In the following the problem of superresolution is re-examined for objects given by primary or secondary continuous sources that are bounded in space and confined to a prescribed two-dimensional or three-dimensional region of space of order $\lambda^2$ or $\lambda^3$, respectively. For brevity, a source so localized will be called "$\lambda$-localized" or localized in a "$\lambda$-region" of space. Presupposing such a source of scattered or diffracted radiation does not introduce any unrealistic prior knowledge but accounts for a property of many scattering or diffracting objects. It can be shown that I) any band-limited spatial-frequency spectrum in the near zone is *inconsistent* with a $\lambda$-localized source [4.24-28], II) given a $\lambda$-localized source, the exterior field is to be represented by waves associated with *complex-valued* spatial frequencies; the magnitude of these waves remains significant in the far zone, even if the real part of the associated spatial frequencies is in excess of the diffraction limit [4.29].

## 4.2 Partial Waves Associated with Complex Spatial Frequencies

Figure 4.1 is to show that, as one might expect, the field in the neighborhood of a field point in the far zone is *locally* approximated by the spatial distribution of a homogeneous plane wave. But the field in the *whole* exterior of a localized source is neither represented nor validly approximated by a linear combination only of

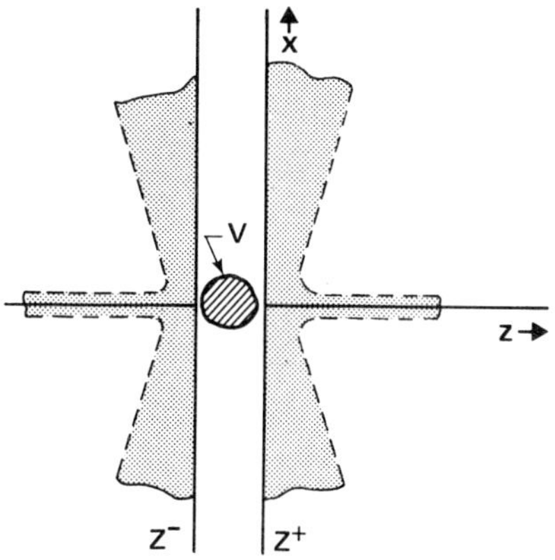


Fig. 4.2. Halfspace representation by plane waves. The approximation to the halfspace field by homogeneous plane waves is inconsistent in the shaded regions with the field of a source localized in the λ-region V

homogeneous plane waves. It can be shown (see Sect.4.4) that a field composed only of homogeneous plane waves fails, in the shaded regions schematically indicated in Fig.4.2, to be a valid approximation of the field of a source confined to the λ-region V. A field composed only of homogeneous plane waves may be a valid approximation to the field of such a source everywhere outside the shaded regions indicated in Fig.4.2. Similarly, the expansion of the exterior field in terms of multipole waves with the multipoles at the origin, which suffices to approximate the exterior field for large radial distances, fails in the regions shown in Fig.4.3 to be a valid approximation to the field of a source confined to the λ-region V.

Real-valued planar spatial frequencies $f_x, f_y > 1/\lambda$ in one of the planes $Z^+$ or $Z^-$ are commonly associated [4.30] with nonuniform-plane waves that move along those planes and are evanescent in the positive or negative z direction, respectively (see Figs.4.2 and 4.4b). Such a waveform suggests that a high-frequency disturbance is not propagated to the far zone but is exponentially attenuated away from the object plane [4.31,32]. However, it is well known [4.29,33] that there are infinitely many bases of plane waves from which to compose planar field distributions given over $Z^+$ or $Z^-$, namely nonuniform plane waves with a nonzero z-component of the wave propagation vector. Any nonuniform plane wave is evanescent in a direction *perpendicular* to the direction of propagation [Ref.4.34, pp.672-79] (see Fig.4.7).

The choice between these different bases of plane waves is one between real and complex-valued planar spatial frequencies; homogeneous plane waves and plane waves that are evanescent in the z direction are associated with *real*-valued planar spatial frequencies. The general nonuniform plane wave is associated with a *complex*-valued planar spatial frequency. Accordingly, wave functions that belong to different real-valued spatial frequencies are well known to be mutually *orthogonal* over a plane, while the wave functions that belong to different complex spatial frequencies are *not* mutually orthogonal. For convenience, one has preferred to disregard complex-valued spatial frequencies and has commonly represented a halfspace field by plane waves given by a line integral along the L-path (see Figs.4.4a and b and 4.5), i.e.,

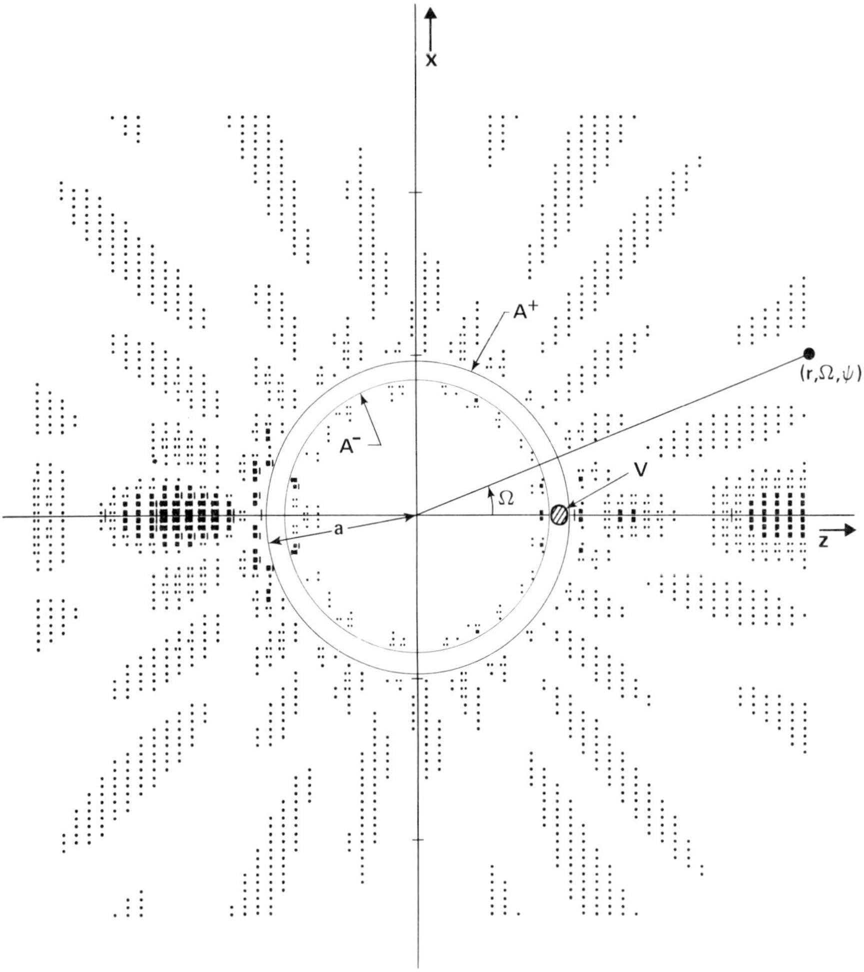


Fig. 4.3. Representation of the exterior field by multipole waves with the multipoles at the origin. Shadings indicate regions where the diffraction-limited field, i.e., the partial sum over multipole waves with real-valued degree $\ell \leq L = 2\pi a/\lambda$ [see (4.3)], fails to be an approximation to the field of a source localized in the $\lambda$-region V. In the shaded regions the relative error of the magnitude of the diffraction-limited field is between 7% and 100%

by a continuous linear combination of plane waves that are either homogeneous or evanescent in the z direction.

The halfspace-field representation by plane waves is uniquely specified by the field distribution over the planes $Z^+$ or $Z^-$. Moreover, a field is *caused* by the action of "point sources" out of which a given source is composed (Huygens' principle). Hence, if the halfspace field to be represented is due to $\lambda$-localized planar or volume sources, we shall hypothesize that any one plane wave is to make a significant contribution to the halfspace field only in those regions of a halfspace where

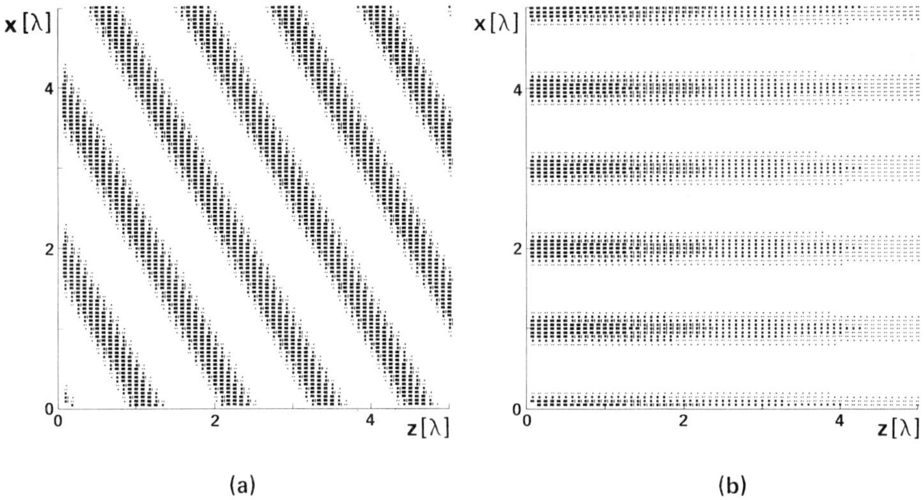


Fig. 4.4a and b. Positive amplitude of plane waves with real-valued planar frequencies. (a) Homogeneous plane wave in a direction of 30° relative to the +z axis. (b) Plane wave evanescent in +z direction which propagates in the direction of 90° relative to the z axis

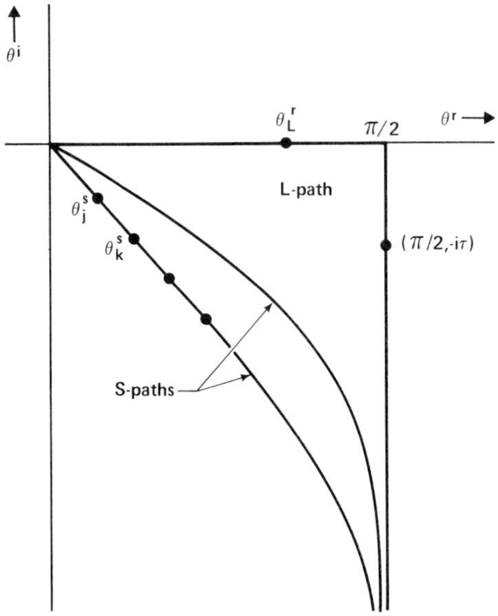


Fig. 4.5. Plane of complex $\theta$. The planar frequencies $f_x$, $f_y$, and the components of the propagation vector $\underline{k}$ of plane waves are related to $\theta$ as given in (4.19, 20, 21)

the field of this wave is causally related to the source. For example, when the source is a point source at the origin the field of any one plane wave is to contribute significantly to the total field only at field points where the phase $\underline{k} \cdot \underline{x}$ [see (4.19-21)] of the plane wave is positive, i.e., to the right of the wave-front $S_0$ (see Fig.4.6) that contains the source point. The *precursor of any plane* wave,

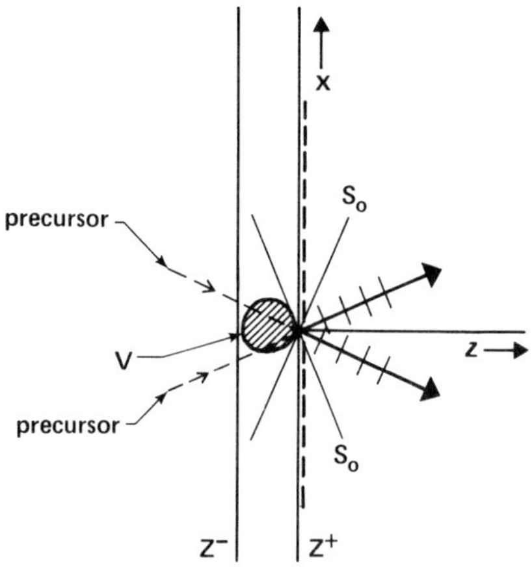


Fig. 4.6. Non-causal precursors of partial plane waves

i.e., that part of the field of a plane wave that moves toward the source point, is to make only an *insignificant* contribution to the halfspace field since it is not causally related to the source point (see note added in proof).

Decomposing the halfspace field in terms of homogeneous plane waves and plane waves that are evanescent in the z direction (see Fig.4.4) does not accord with the above hypothesis. For instance, the field outside and close to the planes $Z^+$ and $Z^-$ is composed significantly of homogeneous and evanescent plane waves whose direction of propagation is at an angle $\theta^r = \pi/2$ relative to the positive z axis. All these plane waves are given by the points on the vertical line from $\theta = (+\pi/2, j0)$ to $(+\pi/2, -j\infty)$ (see Fig.4.5). Only a part of the field of any of these plane waves is causally related to a source point at the origin because presumedly there are no physical provisions to reflect any of these plane waves. Omission of the precursors of the L-path plane waves (see Figs.4.5 and 4.6) can render a halfspace field altogether different from the field to be represented.

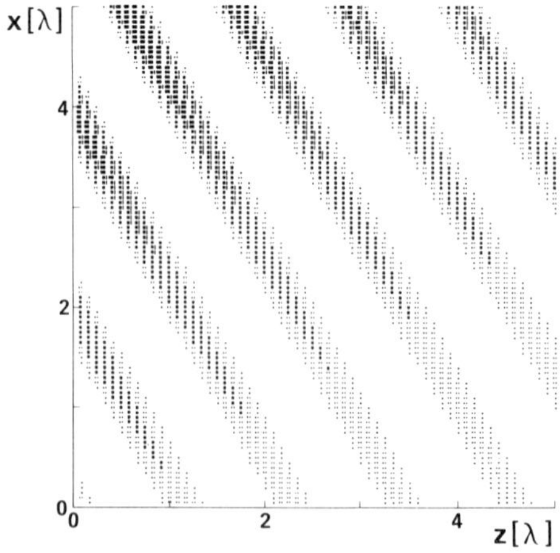


Fig. 4.7. Positive amplitude of a plane wave associated with complex-valued planar spatial frequencies in the direction of propagation of 30° with respect to the +z axis

In contrast, by decomposing the halfspace field in terms of nonuniform plane waves with a nonzero z-component of the wave propagation vector [S-path plane waves, see (4.25), Figs.4.5 and 4.7] we not only fit the field to the given distribution over one of the planes $Z^+$ or $Z^-$ but also meet with the above hypothesis with regard to a causal relation between the halfspace field and a λ-localized source. The field of any of these nonuniform plane waves contributes significantly to the halfspace field only where it is propagated away from the source, for example from a point source at the origin, because of the exponential decay of these plane waves perpendicular to the direction of propagation; the precursors of these waves produce a field the magnitude of which is negligible everywhere in the halfspace, possibly except in the vicinity of a source. A continuous sum of nonuniform plane waves with directions of propagation between 0 and 90 degrees with respect to the z axis provides for an unband-limited representation of the halfspace field (see Sect.4.5).

The field of a source in a λ-region V that does not contain the origin (see Fig. 4.3) may be expanded outside the sphere $A^+$ in terms of multipole waves (see Sect. 4.3.3) with *complex*-valued degree[1] [Ref.4.35, pp.214-224, 279-289]. Figures 4.8 and 4.9 show for comparison multipole waves with real- and complex-valued degree; these wavefields arise from multipoles at the origin. Multipole waves with complex-valued degree can be used to represent the exterior field of a source localized in the λ-region V in accord with a causal relation between the exterior field and this source, whereas any multipole wave with real-valued degree belongs to a (fictitious) multipole at the origin [Ref.4.35, pp.150-154]. Multipole waves with complex degree have a significant order of magnitude both in the near and the far zone (see Fig.4.9).

## 4.3 Representations and Expansions of the EM Field

### 4.3.1 Integral Representations

The EM field at any given field point is represented by volume integrals of the source distribution and surface integrals of the EM field [Ref.4.36, pp.48-56] (Huygens' principle). For scalar, time-harmonic fields Huygens' principle can be put more simply [Ref.4.37, pp.199-200] by using the Green's function that vanishes over the boundary surface. The Green's functions that belong to a plane [Ref.4.38, pp.43-45] and a sphere [Ref.4.35, pp.198-200] are commonly used in diffraction theory. The resulting Fredholm integral equations of the first kind have been inverted numerically to obtain, from far-zone field data, values of the EM field at discrete field points in the near zone of the object [4.39,40], and a discrete distribution in three spatial

---

[1] For our terminology "degree" and "order" see the text following (4.2).

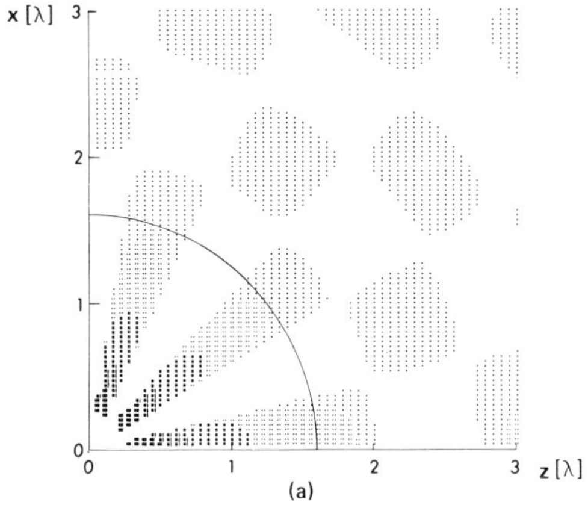


(a)

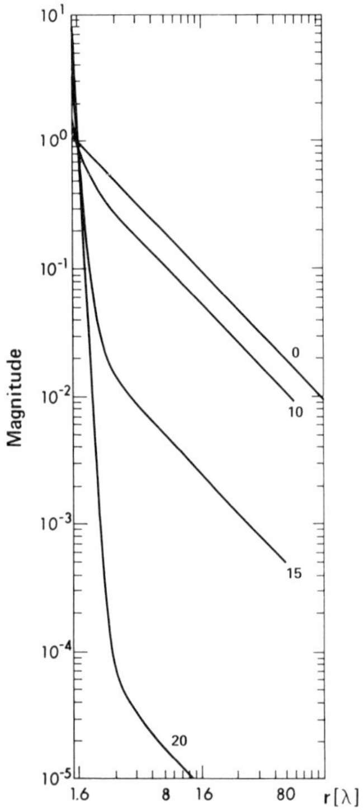


(b)

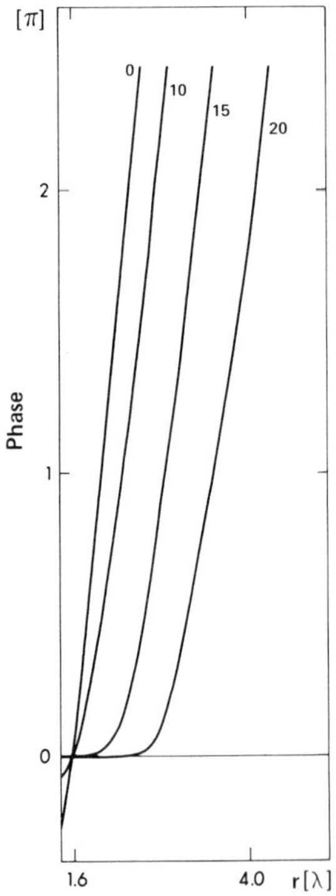


(c)

Fig. 4.8a-c. Multipole waves with real-valued degrees, $\ell \gtreqless ka = 10$.

(a) Ω-dependence of the multipole wave with degree $\ell = 10$. The density plot shows positive values of $\mathrm{Im}\{h_{10}^{(1)}(kr)P_{10}(\cos\Omega)/h_{10}^{(1)}(ka)\}$. The arc indicates $kr = ka = 10$, i.e., $r = 1.6\lambda$.

(b) r-dependence of the magnitude of $h_\ell^{(1)}(kr)/h_\ell^{(1)}(ka)$. The parameter denotes values of $\ell$.

(c) r-dependence of the phase of $h_\ell^{(1)}(kr)/h_\ell^{(1)}(ka)$. The parameter denotes values of $\ell$

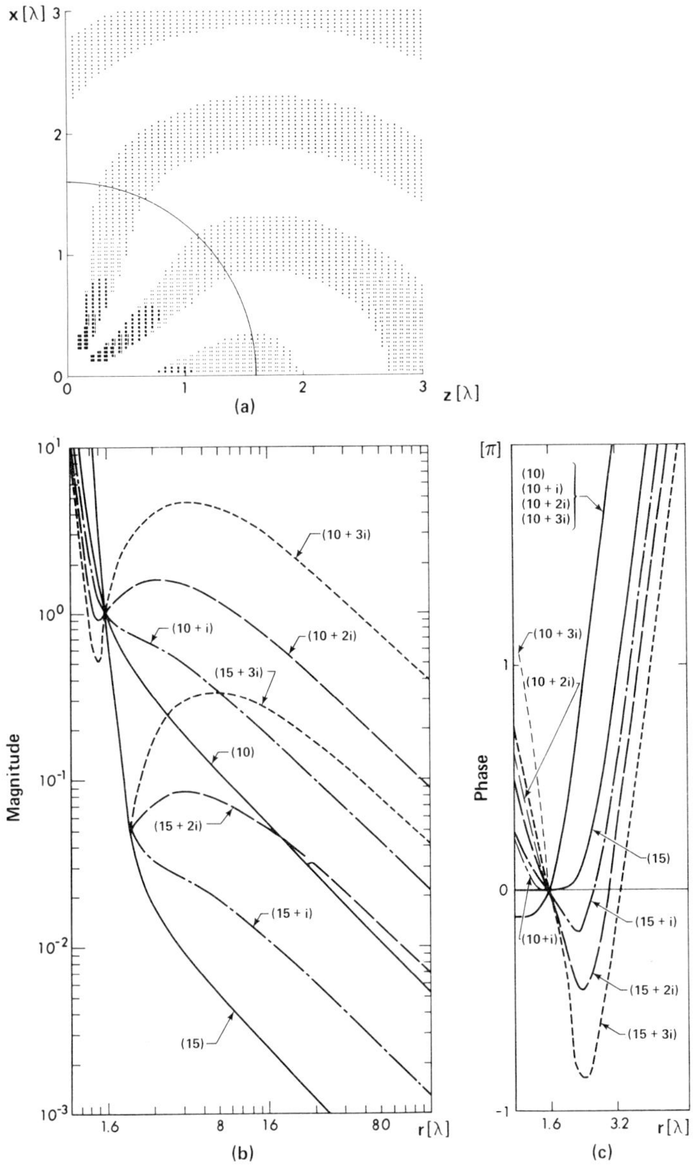


Fig. 4.9a-c. Multipole waves with complex-valued degrees, $Re\{\ell\} \geqq ka = 10$.

(a) $\Omega$-dependence of the multipole wave with degree $\ell = 10+i$. The density plot shows the positive value of $Re\{h^{(1)}_{10+i}(kr)P_{10+i}(-\cos\Omega)/h^{(1)}_{10+i}(ka)\}$. The arc indicates $kr = ka = 10$, i.e., $r = 1.6\lambda$.

(b) r-dependence of the magnitude of $h^{(1)}_{\ell}(kr)/h^{(1)}_{\ell}(ka)$. The parameter denotes values of $\ell$.

(c) r-dependence of the phase of $h^{(1)}_{\ell}(kr)/h^{(1)}_{\ell}(ka)$. The parameter denotes values of $\ell$

dimensions of the scattering source density from Born's approximations [4.41]. Neumann's series solutions have been given for the field inside [4.42] and outside [4.43] a scatterer. Any near-field distribution obtained by numerical inversion is a vector in a signal space with a *finite* dimension; hence, numerical inversion may be used only when the object has, or may approximately be represented by, a finite number of degrees of freedom. The solution obtained with such a discrete method of inversion specifies, for example, the mean value of the refractive index within a volume element [4.41].

### 4.3.2 Partial-Wave Representation of Exterior Field

The outgoing EM field of a time-harmonic source at any field point exterior to the source may be represented by a discrete or a continuous sum of partial waves such that I) every partial wave is a solution to the wave equation in free space, II) the sum (or integral) of partial waves satisfies boundary conditions given over a surface enclosing or excluding all sources [4.44,45] (for example, given over the plane $Z^+$ or $Z^-$, or over the sphere $A^+$ or $A^-$; see Figs.4.2 and 4.3). The physical significance of a partial wave or of some portion of the partial-wave spectrum has sometimes been controversial [4.46], and the physical possibility of superresolution of $\lambda$-localized sources depends again on the proper choice of partial waves for the representation of the field outside such a source.

We now discuss the expansion of the EM field of a localized source (see Figs.4.2 and 4.3) in terms of both multipole and plane waves. If the field arises from a primary source, the scalar quantity $u(\underline{x})$ denotes one component of the vector potential [4.47] [Ref.4.34, pp.704-708] or the Hertzian potential [Ref.4.35, pp.238,268; Ref.4.36, pp.106-108]. Here $\underline{x}=(x,y,z)$ denotes the position vector of a field point. In studying a planar source, e.g., a Gaussian source (see Sects.4.4 and 4.6), $u(\underline{x})$ or $v(\underline{x})$ designate a transverse component of the electric or magnetic field [Ref.4.48, p.129; Ref.4.49, p.268]. For brevity, $u(\underline{x})$ or $v(\underline{x})$ will sometimes be called a "field".

### 4.3.3 Multipole Waves

Let $u(r,\Omega,\psi)$ denote one component of the vector potential [4.45], where $r,\Omega,\psi$ are spherical coordinates of a field point, and let all sources be enclosed by a sphere $A^+$ with radius $r=a$ (see Fig.4.3). Here $u(\underline{x})$ is to satisfy Sommerfeld's radiation condition. The distribution over the sphere $A^+$ of the vector potential due to the interior source distribution is denoted by $u_{A^+}(\Omega,\psi)=u(a,\Omega,\psi)$. It is well known that, for $r>a$, $u(r,\Omega,\psi)$ has the following representation in terms of multipole waves with the multipoles at the origin:

$$u(r,\Omega,\psi) = \sum_{\ell=0}^{\infty} \sum_{m=-\ell}^{+\ell} a_{\ell m} h_\ell^{(1)}(kr) Y_\ell^m(\Omega,\psi) \quad , \tag{4.1}$$

where the coefficients are given by

$$a_{\ell m} = \frac{1}{h_\ell^{(1)}(ka)} \int_0^{2\pi} \int_0^{\pi} u_{A^+}(\Omega',\psi') Y_\ell^{m*}(\Omega',\psi') \sin\Omega' \, d\Omega' d\psi' \quad . \tag{4.2}$$

Our notation follows GOERTZEL and TRALLI [Ref.4.50, p.154]. Here $k = 2\pi/\lambda$, and $Y_\ell^m$ denote spherical harmonics of degree $\ell$ and order m, which are orthogonal over the unit sphere. The asterisk indicates the conjugate complex value. Moreover, $h_\ell^{(1)}$ are spherical Hankel functions of the 1st kind [Ref.4.50, p.157]. The series (4.1) converges absolutely and uniformly for $r > a$ [4.45].

The nodal lines defined by $\mathrm{Re}\{Y_\ell^m(\Omega,\psi)\} = 0$, where Re denotes the real part of the quantity in brackets, define a tesseral grid over the unit sphere [Ref.4.35, p.127]. These nodal lines are equidistantly spaced in $\psi$ direction, while the nodal lines are approximately equidistant in $\Omega$ direction for large values of $\ell$ except close to the two poles, $\Omega = 0$ and $\Omega = \pi$. Thus we can associate with any multipole wave two (instantaneous) spatial frequencies $f_\Omega$, $f_\psi$ in $\Omega$ direction and $\psi$ direction, respectively, on the surface of the sphere $A^+$ given by

$$f_\Omega \simeq \frac{\ell}{2\pi a} \tag{4.3}$$

$$f_\psi = \frac{m}{2\pi a} \quad . \tag{4.4}$$

The coefficient $a_{\ell m}$ given by (4.2) denotes the amplitude of an outgoing multipole wave associated with spatial frequencies $f_\Omega$, $f_\psi$ of the field over the sphere $A^+$. For $\ell = m = 2\pi a/\lambda$ these spatial frequencies exceed the diffraction limit $1/\lambda$. If the distribution of the vector potential over $A^+$ contains "high-spatial-frequency" components, there are nonzero amplitudes $a_{\ell m}$ with $\ell$ greater than $2\pi a/\lambda$. The multipole wave $h_\ell^{(1)} Y_\ell^m$ has an almost constant phase for values of $kr < \ell$ and has outgoing character for $kr > \ell$ (see Fig.4.8). The magnitude of any multipole wave is $O(r^{-(\ell+1)})$ as $r \to 0$ and $O(r^{-1})$ as $r \to \infty$. Thus a multipole wave of large degree decreases rapidly with r in the exterior space close to the sphere $A^+$ (see Fig.4.8) so that a multipole wave with unit magnitude on the sphere $A^+$ has an *insignificant* magnitude in the far zone except for small values of $\ell$; but a small value of $\ell$, i.e., $\ell = 0,1,\ldots$, is associated with high spatial frequencies (4.3,4) if the radius ka of the sphere $A^+$ is a small number, and the monopole wave that belongs to $\ell = 0$, i.e., $h_0^{(1)} Y_0^0 = \exp(ikr)/(ikr)$, varies as $r^{-1}$ in the whole space. It can be shown (see Sect.4.4) that although high-frequency multipole waves (i.e., $\ell$ real-valued $> ka$) are hardly propagated into the far zone, by removing these waves from the multipole-wave spectrum we obtain a field that is at variance with the field of a source localized in a $\lambda$-region of space that does not contain the origin.

A special case of (4.1) is a (TM-mode) point-source dipole in z direction located at $r=r_0$, $\Omega=\Omega_0=0$. The expansion of the vector potential of such a dipole in terms of multipole waves is well known [Ref.4.35, p.145].

$$\frac{e^{ikr}}{ikr}=\begin{cases}\sum\limits_{\ell=0}^{\infty}(2\ell+1)P_\ell(\cos\Omega)j_\ell(kr_0)h_\ell^{(1)}(kr) & , \quad r>r_0 \qquad (4.5)\\ \sum\limits_{\ell=0}^{\infty}(2\ell+1)P_\ell(\cos\Omega)h_\ell^{(1)}(kr_0)j_\ell(kr) & , \quad r<r_0 \quad , \qquad (4.6)\end{cases}$$

where $P_\ell(\cos\Omega)$ is the Legendre polynomial of degree $\ell$, $j_\ell$ are spherical Bessel functions [Ref.4.50, p.157] and

$$R=(r^2+r_0^2-2rr_0\cos\Omega)^{1/2} \quad . \tag{4.7}$$

If the well-known expansion [4.51] for $h_\ell^{(1)}$

$$h_\ell^{(1)}(kr)=\frac{e^{ikr}}{ikr}\sum_{n=0}^{\ell}\frac{b_{\ell n}}{(kr)^n} \quad , \tag{4.8}$$

where

$$b_{\ell n}=(-2)^{-n}\,i^{-n-\ell}\,\frac{(\ell+n)!}{n!(\ell-n)!} \quad , \quad \ell\geqq n \quad , \tag{4.9}$$

is inserted in (4.1), we obtain

$$u(r,\Omega,\psi)=\frac{e^{ikr}}{ikr}\sum_{n=0}^{\infty}\frac{u_n(\Omega,\psi)}{(kr)^n} \quad , \tag{4.10}$$

where

$$u_n(\Omega,\psi)=\sum_{\ell=n}^{\infty}b_{\ell n}\sum_{m=-\ell}^{+\ell}a_{\ell m}Y_\ell^m(\Omega,\psi) \quad , \quad n=0,1,\ldots \quad . \tag{4.11}$$

By (4.11), the "radiation pattern"

$$u_0(\Omega,\psi)=\sum_{\ell=0}^{\infty}b_{\ell 0}\sum_{m=-\ell}^{+\ell}a_{\ell m}Y_\ell^m(\Omega,\psi) \tag{4.12}$$

contains the multipole-wave amplitudes $a_{\ell m}$ of *all* degrees and orders, $\ell=0,1,\ldots$, $-\ell\leqq m\leqq+\ell$, while the angular functions $u_n(\Omega,\psi)$ with larger values of n, which are the amplitudes of the higher-order radial functions $\exp(ikr)/[i(kr)^{n+1}]$, are composed of all multipole-wave amplitudes $a_{\ell m}$ with a degree $\ell$ *in excess* of or equal to n.

The well-known recursion formulas [Ref.4.52, p.194; 4.53] for the angular functions $u_n(\Omega,\psi)$ are obtained by the fact [Ref.4.50, p.153] that the spherical harmonics $Y_\ell^m$ satisfy the equations

$$L^2 Y_\ell^m(\Omega,\psi) = \ell(\ell+1) Y_\ell^m(\Omega,\psi) \quad , \quad \ell = 0,1,\ldots, \tag{4.13}$$

where

$$L^2 = \frac{-1}{\sin\Omega}\left[\frac{\partial}{\partial\Omega}\left(\sin\Omega\frac{\partial}{\partial\Omega}\right)+\frac{1}{\sin\Omega}\frac{\partial^2}{\partial\psi^2}\right] \quad . \tag{4.14}$$

By (4.9)

$$-2i(n+1)b_{\ell,n+1} = [\ell(\ell+1)-n(n+1)]b_{\ell n} \quad , \quad \ell > n \quad , \tag{4.15}$$

and we obtain

$$\begin{aligned}[L^2-n(n+1)]u_n(\Omega,\psi) &= \sum_{\ell=n}^{\infty} [\ell(\ell+1)-n(n+1)]b_{\ell n} \sum_{m=-\ell}^{+\ell} a_{\ell m} Y_\ell^m(\Omega,\psi) \\ &= \sum_{\ell=n+1}^{\infty} [-2i(n+1)]b_{\ell,n+1} \sum_{m=-\ell}^{+\ell} a_{\ell m} Y_\ell^m(\Omega,\psi) \\ &= -2i(n+1)u_{n+1}(\Omega,\psi) \quad , \quad n = 0,1,\ldots \quad . \end{aligned} \tag{4.16}$$

The recursion formulas (4.16) relate the higher-order angular functions $u_n(\Omega,\psi)$, $n = 1,2,\ldots$, *in principle* to the radiation pattern $u_0(\Omega,\psi)$; but it also occurs from (4.11) and (4.16) that the recursion formulas *cannot* be used to reconstruct the near field from far-zone data. The angular function $u_{n+1}(\Omega,\psi)$ is merely another linear combination of the multipole-wave amplitudes $a_{\ell m}$ that enter into $u_n(\Omega,\psi)$ [see (4.11)], and the precision that can be obtained of the value of a higher-order angular function $u_n(\Omega,\psi)$ is determined by the precision with which the higher-order multipole-wave amplitudes $a_{\ell m}$, $\ell \geqq n$, are given to us. Hence, since a multipole wave with real valued degree $\ell > ka$ has an insignificant magnitude in the far zone (see Fig.4.8), a higher-order angular function $u_n(\Omega,\psi)$ cannot be found from far-zone measurements with useful precision. It can be shown (see Sect.4.5), however, that significant information about higher-order multipole-wave amplitudes $a_{\ell m}$ is contained in a far-field distribution, if this distribution is due to field sources confined to $\lambda$-regions of space.

### 4.3.4 Plane Waves

When all sources are in the exterior of a Cartesian halfspace, i.e., to the left of $Z^+$ or to the right of $Z^-$, respectively (see Fig.4.2), the halfspace field can be represented by plane waves [Ref.4.44, pp.733-735; Ref.4.54, pp.11-37]. Here $u(\underline{x})$ denotes one component of the vector potential [Ref.4.34, pp.704-708; 4.47], the Hertzian potential [Ref.4.35, pp.238,268; Ref.4.36, pp.106-108], or one component of the electric or magnetic field [Ref.4.48, p.129; Ref.4.49, p.268]. Both halfspaces have different origins of z coordinates as shown in Fig.4.2. For $z \geqq 0$ ($z \leqq 0$), i.e., for the positive (negative) halfspace, a value of $z = 0$ denotes a point in the plane $Z^+$ ($Z^-$). If the field satisfies Sommerfeld's radiation condition, $u(\underline{x})$ in the positive (negative) halfspace is uniquely specified by the distribution $u_+(x,y)$ ($u_-(x,y)$) in the plane $Z^+$ ($Z^-$)

$$u(\underline{x}) = \int_{-\infty}^{+\infty}\int_{-\infty}^{+\infty} \hat{u}_\pm(f_x,f_y)\exp\left\{+i2\pi\left[xf_x+yf_y+|z|\sqrt{(1/\lambda)^2-f_x^2-f_y^2}\right]\right\}df_x df_y \quad , \tag{4.17}$$

where the planar-spatial-frequency spectra $\hat{u}_\pm$ are given by

$$\hat{u}_\pm(f_x,f_y) = \int_{-\infty}^{+\infty}\int_{-\infty}^{+\infty} u_\pm(x,y)\exp\left[-i2\pi[f_x x+f_y y]\right]dxdy \quad . \tag{4.18}$$

The imaginary part of the square root in (4.17) is greater than or equal to zero. In general, the representation (4.17) is invalid in the region between the planes $Z^+$ and $Z^-$. Any planar spatial frequencies $f_x$, $f_y$ in the plane $Z^+$ ($Z^-$) are associated with a plane wave $\exp(i\underline{k}\cdot\underline{x})$ such that

$$k_x = 2\pi f_x = \frac{2\pi}{\lambda}\sin\theta\,\cos\phi \quad , \tag{4.19}$$

$$k_y = 2\pi f_y = \frac{2\pi}{\lambda}\sin\theta\,\sin\phi \quad , \tag{4.20}$$

$$k_z = \pm\sqrt{\left(\frac{2\pi}{\lambda}\right)^2-k_x^2-k_y^2} = \pm\frac{2\pi}{\lambda}\cos\theta \quad , \tag{4.21}$$

where the positive (negative) square root is taken for $k_z$ in the positive (negative) halfspace. Here $\theta$ and $\phi$ are the angular spherical coordinates of the propagation vector $\underline{k} = (k_x,k_y,k_z)$ of a plane wave. In the following we will refer for convenience only to the positive halfspace and drop the ± label. Spatial frequencies in the plane $Z^+$ in excess of $1/\lambda$ are given, if $\theta$ or $\phi$, or both [Ref.4.55, p.172], assume complex values [see (4.19,20)]. The plane wave $\exp(i\underline{k}\cdot\underline{x})$ with a complex-valued propagation vector given by (4.19,20,21) is a solution of the scalar Helmholtz equation because

$$k_x^2 + k_y^2 + k_z^2 = \left(\frac{2\pi}{\lambda}\right)^2 \quad . \tag{4.22}$$

The Jacobian of the transformation (4.19,20) is equal to $(1/\lambda^2)\sin\theta\,\cos\theta$. Hence, we obtain for the halfspace field a continuous sum of plane waves [4.29]

$$u(\underline{x}) = \int_0^{2\pi} \int_{(0,i0)}^{(\pi/2,-i\infty)} G(\theta,\phi) \exp\left[i\frac{2\pi}{\lambda}(x \sin\theta \cos\phi + y \sin\theta \sin\phi + z \cos\theta)\right]$$

$$\cdot \cos\theta \sin\theta \, d\theta \, d\phi \quad , \tag{4.23}$$

where the angular plane-wave spectrum $G(\theta,\phi)$ is related to the planar-spatial-frequency spectrum $\hat{u}(f_x,f_y)$ by

$$G(\theta,\phi) = \frac{1}{\lambda^2} \hat{u}\left(\frac{1}{\lambda} \sin\theta \cos\phi, \frac{1}{\lambda} \sin\theta \sin\phi\right) \quad . \tag{4.24}$$

In order for the limits of integration in (4.23) to agree with those in (4.17) we take $\phi$ to be a real number within 0 and $2\pi$ and integrate over $\theta$ between $(0,i0)$ and $(\pi/2,-i\infty)$ along the L-path (see Fig.4.5). However, the integration over $\theta$ in (4.23) between $(0,i0)$ and $(\pi/2,-i\infty)$ can, by Cauchy's theorem, be carried out along any other path than the L-path, provided no singular points of the integrand are traversed upon deformation of the path of integration [4.2; Ref.4.29, p.1066]. A more general path of integration in the strip $(0,i0)$, $(\pi/2,i0)$, $(\pi/2,-i\infty)$, $(0,-i\infty)$ of the $\theta$ plane will be called an S-path (see Fig.4.5).

Plane waves given by points on the L-path are either homogeneous (horizontal part of L-path) or evanescent in the +z direction (vertical part of L-path; see Figs.4.4 and 4.5). In contrast, any nonuniform-plane wave given by a complex value of $\theta = (\theta^r, i\theta^i)$ in the strip $(0,i0)$, $(\pi/2,i0)$, $(\pi/2,-i\infty)$, $(0,-i\infty)$ of the $\theta$ plane propagates in the $(\theta^r,\phi)$ direction with a wavelength $\lambda/\cosh\theta^i$. The magnitude of such a plane wave varies exponentially perpendicular to the direction of propagation but remains of order one in the direction of propagation (see Fig.4.7). In the plane $z = 0$ (i.e., in one of the planes $Z^+$ or $Z^-$, respectively) such a nonuniform plane wave has the distribution

$$\exp\left[i\frac{2\pi}{\lambda}(f_x x + f_y y)\right] = \exp\left[i\frac{2\pi}{\lambda}\sin\theta^r \cosh\theta^i (x \cos\phi + y \sin\phi)\right]$$

$$\cdot \exp\left[-\frac{2\pi}{\lambda}\cos\theta^r \sinh\theta^i (x \cos\phi + y \sin\phi)\right] \tag{4.25}$$

[obtained by inserting $\theta = \theta^r + i\theta^i$ into (4.19,20)]. Two *complex*-valued planar spatial frequencies $f_x$, $f_y$ [see (4.19,20)] are associated with a nonuniform plane wave. For the vertical part of the L-path, $\theta = \pi/2 + i\theta^i$, $\theta^i \leqq 0$, we have plane waves that are evanescent in the +z direction with no exponential variation in the x and y direction [see (4.25)] and are associated with real-valued planar spatial frequencies $f_x$, $f_y$ [see (4.19,20)]. Planar spatial frequencies associated with a nonuniform plane wave are *larger* by a factor $\cosh\theta^i$ than planar spatial frequencies associated with a homogeneous plane wave that has the same values of $\theta^r$ and $\phi$ [see (4.25)]. A nonuniform plane wave given by a value of $\theta$ that is not on the L-path is no linear combination of L-path plane waves. Thus, spectra $G(\theta,\phi)$ of a halfspace field taken along the L-path and an S-path between $(0,i0)$ and $(\pi/2,-i\infty)$ of the $\theta$ plane contain different sets of information because they belong to essentially different partial-wave bases [4.56]. The S-path plane waves provide alternative bases for the decomposition of a halfspace field. In Section 4.2, we have given a hypothesis by which an S-path basis is to be chosen in preference to the L-path basis, if the halfspace field to be represented is due to $\lambda$-localized sources.

## 4.4 Band-Limiting at Variance with $\lambda$-Localized Sources

Partial waves associated with real-valued spatial frequencies in excess of $1/\lambda$ (diffraction limit) are essential, if the field represented by (4.1,17, or 4.23) is to be *consistent* with the presence of a source localized in a $\lambda$-region of space. We first discuss some recent results to show that only a whole integral of plane waves, including all high-planar-spatial-frequency components, yields a representation of the halfspace field of a $\lambda$-localized source in consistency with the expansion (4.10); this expansion characterizes the exterior field of a primary source localized within a sphere of given radius about the origin. It follows from (4.10) that the exterior field satisfies the strong radiation condition [Ref.4.48, p.142]

$$\left(\frac{\partial}{\partial r} - ik\right)u = O(r^{-2}) \tag{4.26}$$

for $r\to\infty$, *uniformly* for all directions $\Omega$ and $\psi$. It follows further from (4.10) [4.27] that the curl of the real part $\underline{S}^r$ of the Poynting vector of the exterior field of a bounded source satisfies the condition

$$\nabla \times \underline{S}^r = O(r^{-3}) \tag{4.27}$$

for $r\to\infty$, *uniformly* for all directions $\Omega$ and $\psi$.

The field $v(\underline{x})$ of a planar source in the plane $z=0$ that satisfies Sommerfeld's radiation condition for all $z \geqq 0$ and is absolutely integrable over the plane $z=0$ [4.25]

$$\int_{-\infty}^{+\infty}\int_{-\infty}^{+\infty} |v(x,y,0)|\,dxdy < \infty \tag{4.28}$$

satisfies the strong radiation condition (4.26) at least in a closure of the positive halfspace, $0 \leqq \Omega \leqq \pi/2 - \varepsilon$, $\varepsilon > 0$ [4.25].

Consistency between the field of a $\lambda$-localized source and the halfspace field represented by an integral of plane waves can be judged also by the behavior of the spatial distribution of the Poynting vector in the near zone, for the real part of the Poynting vector denotes the time average of the energy flux [Ref.4.34, pp.625-628], and the time-averaged energy is to move away from such a source along smooth trajectories in the exterior space.

The asymptotic behavior in a Cartesian halfspace of fields represented by homogeneous and inhomogeneous plane waves has been studied by SHERMAN et al. [4.57,58]. SCHMIDT-WEINMAR, BALTES, and RAMSAY [4.24-28] obtained asymptotic expansions with variable bandlimit of the field of the following $\lambda$-localized sources: 1) a Hertzian dipole, 2) a Gaussian source of small cross-section. In the following we present some results on the field of these $\lambda$-localized sources that apply to the asymptotic limit $\rho = (x^2+y^2)^{1/2} \to \infty$, for a fixed value of z. The plane-wave spectra of these fields were integrated along the L-path between (0,i0) and $(\pi/2,-i\tau)$, where the cutoff parameter $\tau$ is related to the upper bandlimit $f_\tau$ of real-valued planar spatial frequencies by $f_\tau = (1/\lambda)\cosh\tau$ (see Fig.4.5). We choose $\tau > 0$ so that halfspace fields are considered that contain *all* homogeneous plane-wave components *plus* those plane-wave components that are evanescent in the z direction and are associated with real-valued planar spatial frequencies between $1/\lambda$ and $(1/\lambda)\cosh\tau$.

If the real-valued spatial frequencies are limited to absolute values between zero and $f_\tau = (1/\lambda)\cosh\tau$, the vector potential $u^\tau(\rho,z)$ of a Hertzian dipole at the origin is given by

$$u^\tau(\rho,z) = \int_0^{\frac{\pi}{2}-i\tau} J_0(k\rho \sin\theta) \exp(ikz \cos\theta) \sin\theta \, d\theta \quad , \quad z > 0 \quad , \tag{4.29}$$

where $k = 2\pi/\lambda$. In the limit $\tau \to \infty$, i.e., with infinite bandwidth of real-valued planar spatial frequencies, (4.29) represents a spherical wave [Ref.4.35, p.242]

$$\lim_{\tau\to\infty} u^\tau(\rho,z) = \frac{\exp(ikr)}{ikr} \quad , \quad z > 0 \quad , \tag{4.30}$$

where

$$r = (\rho^2+z^2)^{1/2} \quad . \tag{4.31}$$

If $v^\tau(\rho,z)$ denotes the x component of the electric field of an axis-symmetric Gaussian planar source [4.32] whose real-valued planar spatial frequencies are limited to absolute values between zero and $f_\tau = (1/\lambda)\cosh\tau$, the halfspace field is given by

$$v^\tau(\rho,z) = \int_0^{\frac{\pi}{2}-i\tau} J_0(k\rho \sin\theta) \exp\left[-\frac{(k\sigma)^2}{2}\sin^2\theta\right] \cdot \exp(ikz \cos\theta) \sin\theta \cos\theta \, d\theta \quad , \quad z \geqq 0 \quad , \tag{4.32}$$

with

$$\lim_{\tau\to\infty} v^\tau(\rho,0) = \frac{\exp\left(-\frac{\rho^2}{2\sigma^2}\right)}{(k\sigma)^2} \quad . \tag{4.33}$$

Similar integrals are obtained [4.28] for the y- and z-components of the electric field and the magnetic field of a planar Gaussian source.

For $z=0$, both unband-limited fields (4.30) and (4.33) satisfy the strong radiation condition[2] (4.26) as $\rho\to\infty$, viz.,

$$\left(\frac{\partial}{\partial r} - ik\right)\frac{\exp(ik\rho)}{ik\rho} = ik\,\frac{\exp(ik\rho)}{(k\rho)^2} \quad , \tag{4.34}$$

$$\left(\frac{\partial}{\partial r} - ik\right)\frac{\exp\left(-\frac{\rho^2}{2\sigma^2}\right)}{(k\sigma)^2} = -\frac{\exp\left(-\frac{\rho^2}{2\sigma^2}\right)}{(k\sigma)^2}\left[\left(\frac{\rho}{\sigma^2}\right)+ik\right] \quad . \tag{4.35}$$

But the integrals (4.29) and (4.32) have the following leading terms when they are asymptotically expanded for $\rho\to\infty$ and fixed values of $z$ and $\tau$,

$$u^\tau(\rho,z) = \frac{\exp(ik\rho)}{ik\rho} + i\left(\frac{2}{\pi}\right)^{1/2}\frac{\cosh^{1/2}\tau}{\sinh\tau}\,\frac{\exp(-kz\sinh\tau)}{(k\rho)^{3/2}} \cdot \cos(k\rho\cosh\tau+\pi/4) + O(\rho^{-5/2}) \quad ,$$

$$(\text{Hertzian dipole},\ \tau>0,\ z>0,\ \rho\to\infty) \quad , \tag{4.36}$$

$$v^\tau(\rho,0) = -\left(\frac{2}{\pi}\right)^{1/2}\frac{\cosh^{1/2}\tau}{(k\rho)^{3/2}}\,\frac{\exp\left[-\frac{(k\sigma\cosh\tau)^2}{2}\right]}{(k\sigma)^2}\cos(k\rho\cosh\tau+\pi/4) + O(\rho^{-5/2}) \quad ,$$

$$(\text{Gaussian source},\ \tau\geq 0,\ \rho\to\infty) \quad . \tag{4.37}$$

As $\rho\to\infty$, the band-limited halfspace fields satisfy radiation conditions of order $O(\rho^{-3/2})$ for fixed $z$, viz.,

$$\left(\frac{\partial}{\partial r} - ik\right)u^\tau(\rho,z) = ik\left\{\frac{\exp(ikr)}{(kr)^2} - \left(\frac{2}{\pi}\right)^{1/2}\frac{\cosh^{1/2}\tau}{\sinh\tau}\,\frac{\exp(-kz\sinh\tau)}{(k\rho)^{3/2}} \cdot[\cosh\tau\sin(k\rho\cosh\tau+\pi/4)+i\cos(k\rho\cosh\tau+\pi/4)]\right\} + O(\rho^{-5/2})$$

$$(\text{Hertzian dipole},\ \tau>0,\ z>0,\ \rho\to\infty) \tag{4.38}$$

---

[2]The unband-limited Gaussian field in the plane $z=0$ satisfies a stronger radiation condition of order $O[\rho\exp(-\rho^2/2\sigma^2)]$, see (4.35).

and

$$\left(\frac{\partial}{\partial r} - ik\right)v^{\tau}(\rho,z) = k\left(\frac{2}{\pi}\right)^{1/2} \frac{\cosh^{1/2}\tau}{(k\rho)^{3/2}} \frac{\exp\left[-\frac{(k\sigma \cosh\tau)^2}{2}\right]}{(k\sigma)^2} [\cosh\tau \sin(k\rho \cosh\tau+\pi/4)$$

$$+i \cos(k\rho \cosh\tau+\pi/4)] + O(\rho^{-5/2})$$

$$(\text{Gaussian source, } \tau \geqq 0,\ z=0,\ \rho\to\infty) \quad . \tag{4.39}$$

Equations (4.38) and (4.39) show that for the band-limited halfspace fields to satisfy approximately a radiation condition of order $O(\rho^{-2})$ for $\rho\to\infty$ [4.24,25]

$$\exp(-kz \sinh\tau) \frac{\cosh^{3/2}\tau}{\sinh\tau} (k\rho)^{1/2} \ll 1$$

$$(\text{Hertzian dipole, } \tau>0,\ z>0,\ \rho\to\infty) \quad , \tag{4.40}$$

$$\exp\left[-\frac{(k\sigma \cosh\tau)^2}{2}\right] \cosh^{1/2}\tau\ (k\rho)^{1/2} \ll 1$$

$$(\text{Gaussian source, } \tau \geqq 0,\ z=0,\ \rho\to\infty) \quad . \tag{4.41}$$

Equations (4.40,41) imply the following dependence of the upper bandlimit $f_\tau$ of planar spatial frequencies on z and $\sigma$, respectively, as $\rho\to\infty$,

$$f_\tau = \frac{\cosh\tau}{\lambda} \sim \frac{1}{2\pi z}\left[\frac{\ln(k\rho)}{2} + \text{const.}\right]$$

$$(\text{Hertzian dipole, } z>0,\ \rho\to\infty) \quad , \tag{4.42}$$

$$f_\tau = \frac{\cosh\tau}{\lambda} \sim \frac{1}{2\pi\sigma} [\ln(k\rho)+\text{const.}]^{1/2}$$

$$(\text{Gaussian source, } \sigma>0,\ z=0,\ \rho\to\infty) \quad . \tag{4.43}$$

This shows that no finite value of the upper bandlimit $f_\tau$ of planar spatial frequencies suffices to make the fields $u^\tau(\rho,z)$ and $v^\tau(\rho,z)$ satisfy approximately the strong radiation condition (4.26) (see Fig.4.10) as $\rho\to\infty$ with fixed small values of $z>0$ (Hertzian dipole) or $\sigma>0$ (Gaussian beam). Equivalent results have been obtained [4.27] concerning the asymptotic behavior of the Poynting vector [see (4.27)] for the band-limited EM field of a Hertzian dipole.

Nonuniform plane waves are thus seen to be indispensable to approximate the halfspace field of a $\lambda$-localized source for asymptotically large off-axial distances from the source. Nonuniform plane waves are also indispensable to obtain a proper behavior of the energy flux of the field of such a source. The removal from the plane-wave spectrum of any of the nonuniform-plane-wave component causes the time-averaged

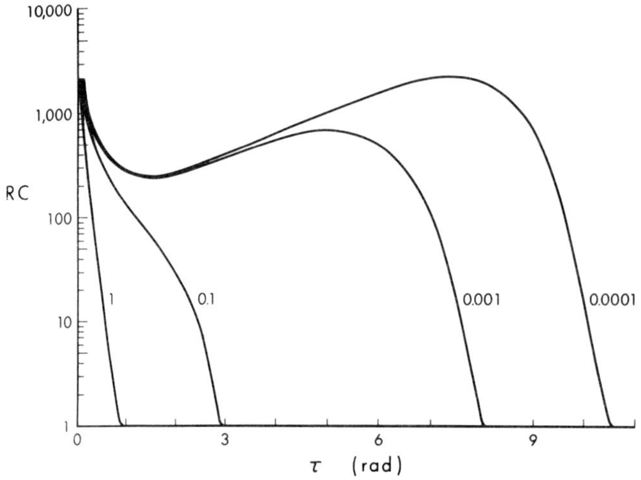


Fig. 4.10. Normalized radiation condition

$$RC = |r^2 k(\frac{\partial}{\partial r} - ik)u^\tau(\rho,z)|$$

for a band-limited point-source dipole field $u^\tau(\rho,z)$, vs the cutoff parameter $\tau$ of evanescent waves (see Fig.4.5). The lateral distance $\rho$ of field points is 10,000 wavelengths. Parameter: the longitudinal distance z of field points in wavelengths. For the unband-limited field, RC = 1

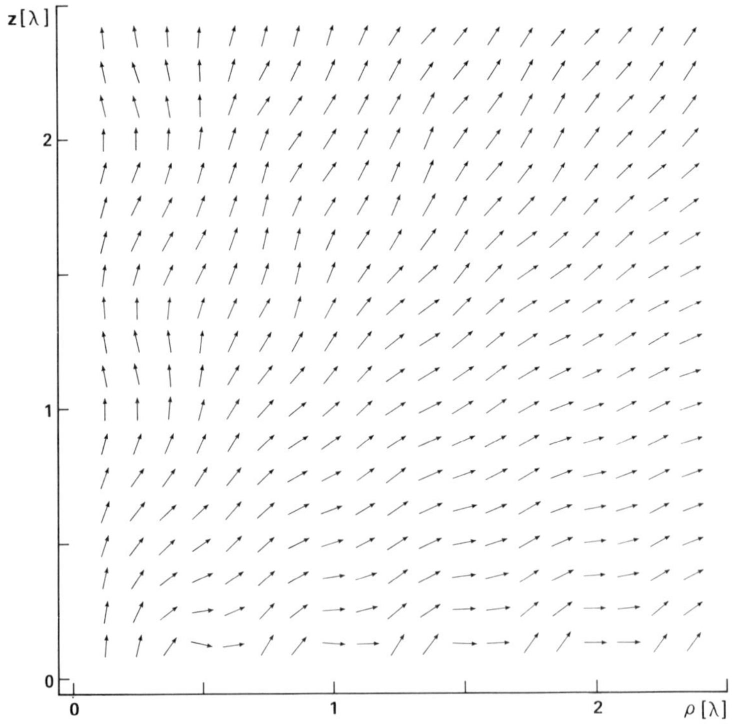


Fig. 4.11. Meandering along smooth trajectories of the time-averaged Poynting vector of the band-limited EM field of a λ-localized planar source. The upper bandlimit of planar spatial frequencies is equal to 1/λ. The unband-limited EM field in the plane z = 0 is Gaussianly distributed [see (4.32,33)] with $\sigma = \lambda/8$

energy flux of the field outside a λ-localized source to meander [4.28] about smooth trajectories (see Fig.4.11). Hence, any set of plane waves short of the *entire* plane-wave spectrum results in a halfspace field that even under far-field conditions is *inconsistent* with the field of a λ-localized source.

An inconsistency between a band-limited wave field and a source of small spatial extent is borne out furthermore by the expansion in terms of multipole waves of the vector potential of a point-source dipole located at $r = r_0 \neq 0$, $\Omega = \Omega_0 = 0$. The band-limited series of multipole waves [see (4.5,6)]

$$u^L(r,\Omega) = \sum_{\ell=0}^{L} (2\ell+1) j_\ell(kr_0) h_\ell^{(1)}(kr) P_\ell(\cos\Omega) \quad , \quad r > r_0 \quad , \tag{4.44}$$

$$= \sum_{\ell=0}^{L} (2\ell+1) h_\ell^{(1)}(kr_0) j_\ell(kr) P_\ell(\cos\Omega) \quad , \quad r < r_0 \tag{4.45}$$

are not in conflict with the strong radiation condition (4.26) as $r \to \infty$. Every term of (4.44) satisfies (4.26), and the recursion formulas (4.16) are satisfied with any finite or infinite value of the upper limit L in (4.44,45); this agrees with the view [4.45] that each term of (4.44) expresses the field of a multipole of degree $\ell$

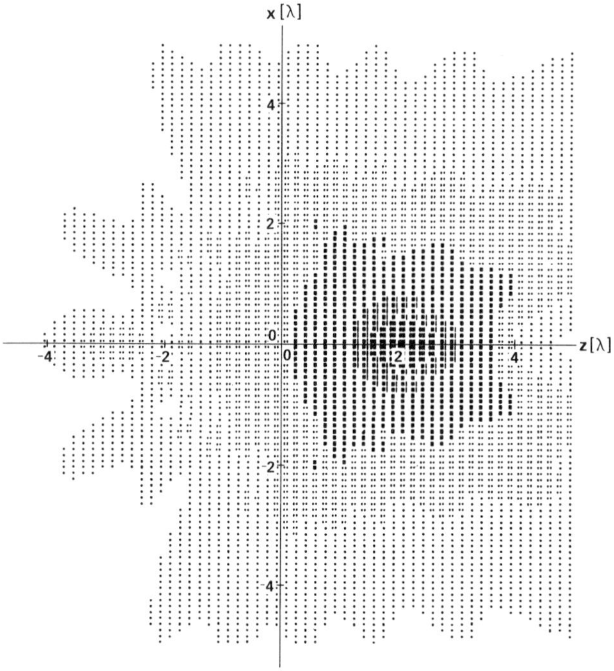


Fig. 4.12. Density plot of the magnitude of the diffraction-limited multipole-wave field of a point-source dipole located at $r=r_0=2\lambda$, $\Omega=\Omega_0=0$. The upper bandlimit L [see (4.44,45)] is given the integer value 13 so that the spatial frequency $f_\Omega$ in $\Omega$-direction [see (4.3)] over the surface $r=r_0$ has the value $1/\lambda$

at the origin. However, if the upper limit L of the series (4.44,45) is given, e.g., the integer value $L = 2\pi r_0/\lambda$ so that $u^L(r,\Omega)$ is band-limited at a spatial frequency $f_\Omega$ [see (4.3)] on the sphere with radius $r = r_0$ given by $f_\Omega \simeq L/2\pi r_0 = 1/\lambda$, we obtain a field the magnitude of which is distributed as shown in Fig.4.12; clearly, this diffraction-limited field is at variance with a source in a $\lambda$-region with the radial coordinate $r = r_0 \neq 0$ (see also the distribution of the relative error of the magnitude of the field shown for this case in Fig.4.3). The field represented by (4.44,45) is inconsistent with a source localized in a $\lambda$-region at $r = r_0 \neq 0$, unless the upper limit L in (4.44,45) is infinite. The exterior neighborhood of the sphere $A^+$ (see Fig.4.3), i.e., $r \to r_0+$, is the region where the highest values of the upper limit L in (4.44) are required to make the exterior field consistent with the field of a source localized in a $\lambda$-region at $r = r_0 \neq 0$; this is analogous to the results obtained when the plane-wave spectrum of the halfspace field of such a source is band-limited.

## 4.5 High-Frequency Information in the Far Zone Given a $\lambda$-Localized Source

The results presented in Sections 4.1.1 and 4.4 indicate how a solution to the problem of superresolution may be found, if the object field arises from $\lambda$-localized sources. Information on the real-valued high spatial frequencies is hardly transmitted to the far zone because the associated partial waves are heavily damped (see Figs.4.4 and 4.8). However, the information associated with the real-valued low spatial frequencies, which is transmitted to the far zone, e.g., via the homogeneous plane waves, is inconsistent with a priori information, if $\lambda$-localized sources are present. This inconsistency can be removed, if the far-zone information is evaluated subject to the *condition* that such sources are given. For instance, when a finite number of point sources are given or when a continuous source distribution can be approximated by a finite number of point sources, a method of evaluation is well known [4.16,40,41] (see Sects.4.1.1 and 4.3.1). Then the spatial-frequency spectra to be found from far-zone data are given by a linear combination of spectra [see (4.2,5,18)] each one basically of the *known* form

$$a_\ell = (2\ell+1) j_\ell(kr_0) \quad , \quad \ell = 0,1,\ldots \quad \text{(point source at } r_0 \neq 0,\ \Omega_0 = 0) \quad , \tag{4.46}$$

$$\hat{u}(f_\rho) = \frac{\lambda}{\left(\left(\frac{1}{\lambda}\right)^2 - f_\rho^2\right)^{1/2}} \quad , \quad 0 \leqq f_\rho < \infty \quad \text{(point source at the origin, [4.29])} \quad . \tag{4.47}$$

In (4.47),

$$f_\rho = \frac{\sin\theta}{\lambda} \quad . \tag{4.48}$$

A fast computer algorithm to solve the inverse problem given λ-localized sources of scattering has been obtained [4.40,41].

The field of a λ-localized source is inconsistent with a band-limited wave field. Consistency between the field reconstructed from far-zone data and the prior knowledge that λ-localized sources are given is obtained most directly by expanding the exterior field in terms of partial waves with *complex*-valued spatial frequencies. Furthermore, the field of a nonuniform plane wave associated with complex-valued planar frequencies is by hypothesis causally related to a λ-localized source (see Sect.4.2). A method of reconstructing the near field of such a source using nonuniform plane waves will be discussed in Section 4.6. An analogous method that would use multipole waves with complex-valued degree (see Fig.4.9), is not known to date, but waves of this type have already been used by SOMMERFELD [Ref.4.35, pp.214-224, 279-289].

## 4.6 λ-Localized Sources Reconstructed from Far-Zone Data

If the halfspace field, $z > 0$, is represented by nonuniform plane waves with a positive direction of propagation, which in the θ plane are given by points on an S-path (see Fig.4.5), each partial wave is propagated through the halfspace without attenuation in the direction of propagation. These complex-planar-frequency waves [see (4.19,20), and Fig.4.7] retain a significant magnitude in the far zone; hence, the task of specifying a spectrum of S-path plane waves from far-zone data is *not* one of extrapolating or continuing analytically any given data but of *interpolating* the far-field data in terms of the expansion (4.23). The interpolation has, however, to be done in agreement with the a priori *analyticity* of the spectrum $G(\theta,\phi)$ [see (4.24)] of S-path plane waves. Unfortunately, an asymptotic representation of the far field that takes plane waves only from the neighborhood of a finite number of points of the S-path does not exist (SCHMIDT-WEINMAR et al. [4.56]).

An algorithm to determine the spectrum function $G(\theta,\phi)$ [see (4.23)] along an S-path in the complex θ plane (see Fig.4.5), from the radiation pattern $u_0(\Omega,\psi)$ [see (4.10)], has been given by SCHMIDT-WEINMAR [4.29]. For large values of r, $z > 0$, the Fredholm integral equation of the first kind

$$\frac{\exp(ikr)}{ikr} u_0(\Omega,\psi) \simeq \int_0^{2\pi} \int_{\text{S-path}} G(\theta,\phi) \exp\left\{i\frac{2\pi}{\lambda} r[\sin\Omega \sin\theta \cos(\phi-\psi) + \cos\Omega \cos\theta]\right\} \cos\theta \sin\theta \, d\theta \, d\phi \tag{4.49}$$

can be solved for $G(\theta,\phi)$ recursively using the fact (see Fig.4.7) that an S-path plane wave is evanescent perpendicular to the direction of propagation. For large values of r, significant contributions to the integral (4.49) are hence made only by those S-path plane waves whose colatitude $\theta^r$ of direction of propagation is smaller than or equal to $\Omega+\varepsilon^S$, where $\varepsilon^S \to 0$ when $r\to\infty$. We can thus determine $G(0,\phi) \equiv G(0)$ from $u_0(0,\psi) = u_0(0)$ by computing

$$\frac{\exp(ikr)}{ikr}\, u_0(0) \cong 2\pi G(0) \int_{(0,i0)}^{\theta_1^S} \exp\left(i\frac{2\pi}{\lambda} r \cos\theta\right) \cos\theta \sin\theta \, d\theta \quad , \tag{4.50}$$

where $\theta_1^S$ (see Fig.4.5) denotes a first value of the upper limit of the line integral along the S-path to be chosen in such a way that the remainder of the complete integral (4.49), makes an insignificant contribution to $[\exp(ikr)/ikr]u_0(0)$ as $r\to\infty$. Since $\theta_1^S \to 0$ as $r\to\infty$, we obtain a valid approximation to the integral by replacing $G(\theta,\phi)$ with a mean value $G(0,\phi) = G(0)$, which can be taken out of the double integral (4.49). Then the value of the remaining integral to the right of $G(0)$ [see (4.50)] can be computed. However, since any S-path plane wave contributes to the field at points with $\Omega > \theta^r$ with a magnitude that grows exponentially with the radial distance r of the field points the contributions to the field at $(r,\Omega,\psi)$ with $\Omega > 0$ are given by two terms:

$$\begin{aligned}\frac{\exp(ikr)}{ikr}\, u_0(\Omega,\psi) \cong &\int_0^{2\pi} \int_{(0,i0)}^{\theta^S(\Omega)-\delta^S} G(\theta,\phi) \exp(iP) \cos\theta \sin\theta \, d\theta \, d\phi \\ &+ \int_0^{2\pi} \int_{\theta^S(\Omega)-\delta^S}^{\theta^S(\Omega)+\varepsilon^S} \ldots \, d\theta \, d\phi \quad ,\end{aligned} \tag{4.51}$$

where $P = P(r,\Omega,\psi,\theta,\phi) = 2\pi r\lambda^{-1}[\sin\Omega \sin\theta \cos(\phi-\psi)+\cos\Omega \cos\theta]$ and where $\theta^S(\Omega)$ is that value of $\theta$ on the S-path whose real part is equal to $\Omega$. Here $\delta^S$ may be any small (complex) number depending on the number of increments $\theta_j^S$, $\theta_k^S$ ... (see Fig.4.5) one chooses for measuring in the $\theta^r$ direction. Moreover $\varepsilon^S \to 0$ as $r\to\infty$. The first term of (4.51) represents the contribution to the field at $(r,\Omega,\psi)$ by S-path plane waves that are propagated in directions with a colatitude $\theta^r < \Omega$, while the second term expresses the continuous sum of all S-path plane waves with a colatitude $\theta^r$ of propagation approximately equal to $\Omega$. The spectrum function $G(\theta,\phi)$ in the first term of (4.51) is known from measurements at smaller values of $\Omega$. We thus recursively find $G(\theta,\phi)$ along an S-path as the solution of a simple and well-conditioned system of simultaneous linear equations which relates values of the radiation pattern with fixed $\Omega$ to values of the S-path plane-wave spectrum with fixed $\theta$ [4.29]

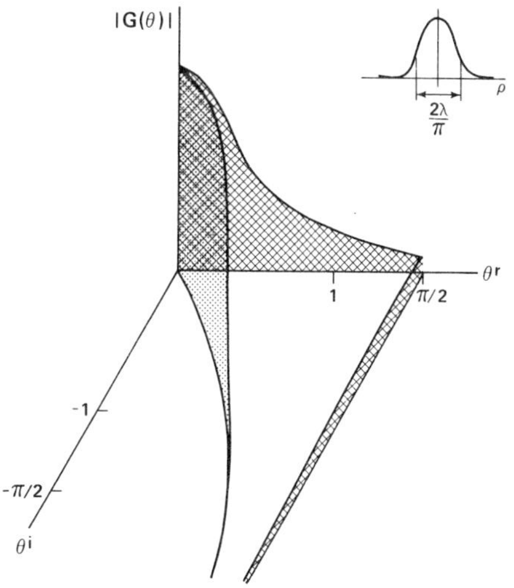


Fig. 4.13. Magnitude of the angular plane-wave spectrum of a Gaussian source in the θ plane. The standard deviation σ of the field in the plane z = 0 is given the value $\lambda/(\sqrt{2}\pi)$ [see (4.53)]

$$\frac{\exp(ikr)}{ikr}\, u_0(\Omega,\psi) - \int_0^{2\pi}\int_{(0,i0)}^{\theta^s(\Omega)-\delta^s} \ldots\, d\theta\, d\phi$$

$$= \int_0^{2\pi} G(\theta_s(\Omega),\phi) \int_{\theta^s(\Omega)-\delta^s}^{\theta^s(\Omega)+\varepsilon^s} \exp(iP)\cos\theta\,\sin\theta\, d\theta\, d\phi \quad , \quad 0 \leqq \psi < 2\pi,\ 0 \leqq \phi < 2\pi \quad . \tag{4.52}$$

All integrals occurring in this algorithm can be computed with high accuracy. These computations may require extended *precision* during the *intermediate* steps of computations because of the exponential growth of the integrand with increasing value of r. An extended precision arithmetic package and library with Fortran precompiler [4.59] may be used. Fortunately, extended precision required for intermediate computations does not always mean that unfeasible *accuracy* is required of the data $u_0(\Omega,\psi)$, as is suggested by the following example.

Figure 4.13 shows the magnitude of the angular plane-wave spectrum $G(\theta)$ of the field $v(\rho,z)$ of a planar Gaussian source in the θ plane

$$G(\theta) = \frac{1}{2\pi}\exp\left[-\frac{(k\sigma)^2}{2}\sin^2\theta\right] \tag{4.53}$$

[see (4.32,33)]. $G(\theta)$ does not depend on $\phi$ because this source is symmetric about the z axis. For $z > 0$, the field is given by

$$v(\rho,z) = \int_{(0,i0)}^{(\pi/2,-i\infty)} J_0(\rho\sin\theta)\exp\left[-\frac{(k\sigma)^2}{2}\sin^2\theta\right]\exp(ikz\cos\theta)\sin\theta\cos\theta\, d\theta . \tag{4.54}$$

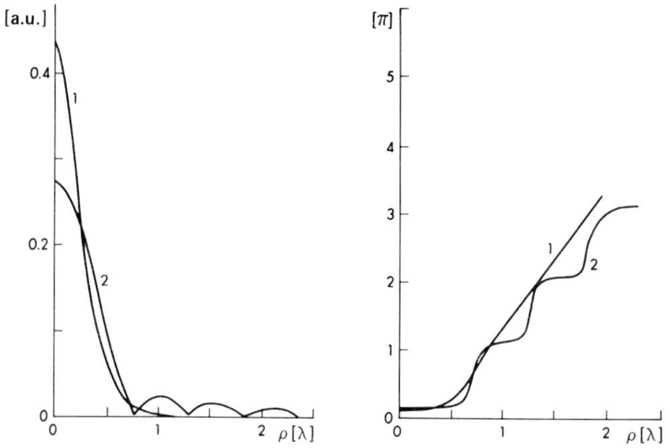


Fig. 4.14. Magnitude and phase of the near field of a Gaussian source in the plane $z = \lambda/10$ reconstructed by plane waves with directions of propagation between zero and 75° with respect to the z axis. The standard deviation $\sigma$ of the field in the plane $z = 0$ [see(4.32,33)] is given the value $\sigma = \lambda/(\sqrt{2}\pi)$

(1) Plane waves with complex-valued planar frequencies $f_x$, $f_y$ [see (4.19,20)]; S-path plane waves. The field reconstructed with S-path plane waves matches the exact field within 0.1%.

(2) Plane waves with real-valued planar frequencies $f_x$, $f_y$ [see (4.19,20)]; homogeneous plane waves

The path of integration in (4.54) is chosen to be the S-path given by

$$\cos\theta^r \cosh\theta^i = 1 \quad , \tag{4.55}$$

i.e., the S-path with the slope -1 at the origin (see Fig.4.5).

We now study the quality of the *reconstruction* of a planar Gaussian source of subwavelength width using the angular plane-wave spectrum along an S-path. First, we compare the near field obtained with S-path plane waves to the field obtained with homogeneous plane waves which have the same directions $\theta^r$ of propagation with respect to the z axis. $0 \leqq \theta^r \leqq \theta_L^r$, where $\theta_L^r = 75°$ (see Fig.4.5). The near field is computed in the plane $z = \lambda/10$ by (4.54) with the upper limit of integration equal to $[\theta_L^r, -i\cosh^{-1}(1/\cos\theta_L^r)]$ for integrating along the S-path, and with the upper limit of integration equal to $(\theta_L^r, i0)$ for integrating along the L-path. The results (see Fig.4.14) show that—for the given example—the continuous sum of S-path plane waves with directions of propagation between zero and 75° with respect to the z axis reconstructs the field close to the plane $z = 0$ extremely well, whereas the continuous sum of homogeneous plane waves with the same directions of propagation deviates significantly from the exact field. The field reconstructed by the homogeneous plane waves is degraded in the way typical for a band-limited spectrum, whereas the field reconstructed by S-path plane waves shows no such degradation.

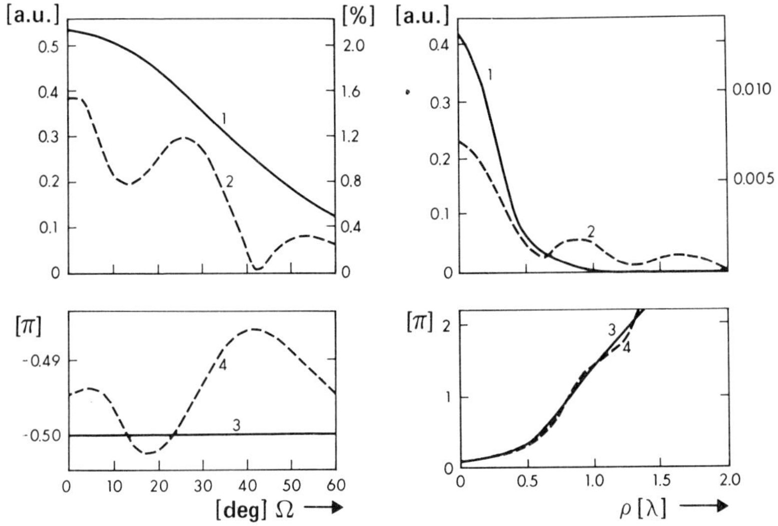


Fig. 4.15. Near-field reconstruction with S-path plane waves from erroneous far-field data. A field is considered that is Gaussianly distributed in a plane $z = 0$ with a standard deviation $\sigma$ of given value $\lambda/(\sqrt{2}\pi)$ [see (4.32,33)].
Left part: Errors of measurement introduced in the far field of a Gaussian source, $r = 300\lambda$. Magnitude (1), phase (3) (scale on lhs) of field without errors, % error (2) introduced in magnitude of field (scale on rhs), and phase (4) of erroneous field.
Right part: Near field reconstructed from the exact and the erroneous far-field data of a Gaussian source in the plane $z = \lambda/10$ as a function of the distance $\rho$ from z axis. Number markings as in left part, except (2) which shows here the magnitude of the errors (scale on rhs)

The near-field distribution of this subwavelength planar Gaussian source reconstructed with S-path plane waves is stable against computer-simulated errors of measurement (noise) in the far-zone distribution [4.25]

$$u(r,\Omega) \simeq \frac{1}{2\pi} \frac{\exp(ikr)}{ikr} \cos\Omega \exp\left[-\frac{(\sigma k)^2}{2} \sin^2\Omega\right] \qquad (4.56)$$

(SCHMIDT-WEINMAR et al. [4.56]). A number of pulsed-shaped model errors distributed about certain mean values of $\Omega$, each given random weights and a standard deviation of a few degrees, were added to the amplitude (4.56) of the far field. Then the S-path plane-wave spectrum was found that belongs to the so degraded far-zone data and, finally, the field in the plane $z = \lambda/10$ was computed from the degraded S-path plane-wave spectrum using (4.54). The results shown in Fig.4.15 are encouraging: errors in the far zone and the near zone have the *same magnitude*. Virtually the same results are obtained for the reconstruction in the plane $z = 0$.

## 4.7 Measurement of Phase and Magnitude of the Optical Radiation Pattern

Since we can indeed find, from the radiation pattern of a $\lambda$-localized source, information on complex-valued frequencies whose real part extends to absolute values greater than the diffraction limit, and since thereby the data have to be measured only with normal (e.g., 1%) precision, the question arises whether a measurement with normal precision of the phase and the magnitude of the radiation pattern of such a source is feasible with optical time-frequencies.

The image of an optical radiation source obtained with practical optical instruments displays only the intensity (squared modulus) of the field. Such an image is degraded by aberrations and diffraction. Even if all aberrations were known to us with sufficient precision, for spatial superresolution the image would have to be processed numerically in order to find the spatial distribution of the phase of the field and to correct for diffraction at the instrument aperture. For simplicity, data acquisition for superresolution is considered in the following without regard to imaging.

The problem of measuring the phase and the modulus of the optical radiation pattern $u_0(\Omega,\psi)$ [see (4.12)] has three components (SCHMIDT-WEINMAR [4.60], see Fig.4.16).

I) Only data of the spatial distribution of the squared modulus of the field are acquired by a photo-detector whereas the spatial distribution of the phase of the field has to be retrieved from several suitable sets of data of the modulus (see Chap.2). It is well known that both optical holography [Ref.4.38, p.200] and interferometry [Ref.4.61, p.36] yield data only of the *difference* of the phase of the spatial distribution functions of the object and the reference wave [see also (4.58, 59)]. Hence, these methods are not immediately useful for our purpose since the spatial distribution of a reference wave is in practice not known to us beyond the diffraction limit.

II) The measurement of the angular distribution of the *phase* of $u_0(\Omega,\psi)$ requires the definition of the surface C of a sphere in the far-zone region of measurement (see Fig.4.17) to within a small fraction $\nu(\cong 10^{-2})$ of the optical wavelength $\lambda$ ($\cong 5.10^{-5}$ cm). For example, with a diffracting object of the linear dimension $\delta \cong 5.10^{-4}$ cm, the diameter D of the surface C, which has to be large in comparison to $\delta^2/\lambda$ to maintain far-field conditions, is of order 1 cm. Hence, a typical *relative accuracy* is required for measuring the radial coordinates of the points of the surface C of the order

$$\frac{\nu\lambda}{D} \sim 10^{-7} \quad . \tag{4.57}$$

III) Since the optical wavelength is small compared to most practical scales of measurement, *optical interferometry* appears to be the only feasible method for measuring the radial coordinates of the surface C. This method suffers, however, from

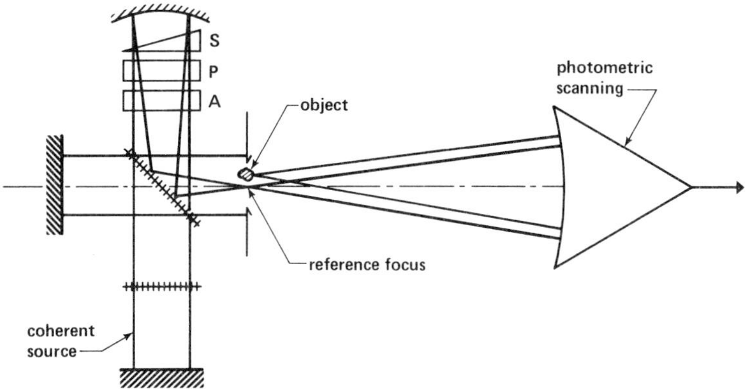


Fig. 4.16. Far-field interferometer with shifted reference wave to measure the angular distribution $u_0(\Omega,\psi)$ of the radiation pattern

the problem mentioned in I) above that without introducing additional information we obtain values only for the relative, but not for the absolute phase.

We now give a brief outline of a method [4.60,62] proposed to solve this particular problem: *far-field interferometry with a shifted reference wave.*

At any far-zone point $(r,\Omega,\psi)$, the measured squared modulus of the field obtained by mixing the object wave $u(\underline{x}) \cong [\exp(ikr)/ikr]u_0(\Omega,\psi) = [\exp(ikr)/ikr] \cdot |u_0(\Omega,\psi)|\exp[id_0(\Omega,\psi)]$ with the field $v(\underline{x}) \cong [\exp(ikr)/ikr]v_0(\Omega,\psi) =[\exp(ikr)/ikr]|v_0(\Omega,\psi)|\exp[ie_0(\Omega,\psi)]$ of a suitable reference focus (Fig.4.16) is given by

$$|u+v|^2 = |v|^2|1+\frac{u}{v}|^2 \quad . \tag{4.58}$$

The spatial distribution of the modulus $|v|$ of the reference wave can be measured without difficulty. To determine the *ratio* u/v, the measurement of $|u+v|$ is repeated twice with a suitably varied value of the reference amplitude v, obtained for example by varying the phase shift at P (see Fig.4.16). This yields the ratio of the radiation patterns of object and reference wave [4.62]

$$\frac{u}{v} \cong \frac{u_0(\Omega,\psi)}{v_0(\Omega,\psi)} \tag{4.59}$$

independent of r. The remaining problem is the elimination of $v_0(\Omega,\psi)$ from the data u/v. Now, $|v_0|$ can be measured without difficulty; hence, the problem is to eliminate the angular distribution $e_0(\Omega,\psi)$ of the phase of the radiation pattern of the reference beam. This can be done by shifting the reference beam in space either translationally or rotationally (see Fig.4.17) without otherwise changing its spatial distribution. Optical elements S can be inserted in the reference beam (see Fig.4.16); or the diffracting element in the reference beam, for example an aperture, may be

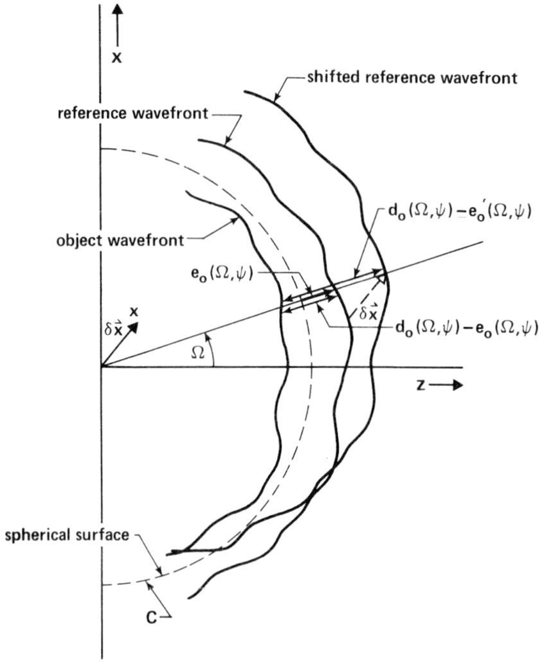


Fig. 4.17. Shifting of reference wavefront to measure the phase $e_0(\Omega,\psi)$ (single-headed arrow) of the radiation pattern of the reference field. $d_0(\Omega,\psi)$ is the phase of the radiation pattern of the object field. The double-headed arrow indicates a quantity that is measured by the photodetector at the point $(r,\Omega,\psi)$. $\delta\underline{x}$ denotes the shifting vector. The prime denotes a quantity after shifting

translated to accomplish the shifting required of the reference beam. The measurement of u/v (4.59) is repeated with the reference beam shifted but fixed object wave. Any shifting is suitable that produces a known linear transformation of the spatial distribution of the reference beam [4.60]. In this way we can provide the precision required to define the radial coordinate of the surface C [see (4.57)].

If, for example, the reference beam is tilted about the origin by angular increments $\delta\Omega$, $\delta\psi$, one obtains the derivatives $\partial e_0/\partial\Omega$ and $\partial e_0/\partial\psi$ from the ratio of the data before (without prime) and after shifting (with prime)

$$\text{data} = \frac{u/v}{u/v'} \cong \left|\frac{v_0}{v_0'}\right| \exp\{i[e_0'(\Omega,\psi)-e_0(\Omega,\psi)]\} \quad , \tag{4.60}$$

where

$$e_0'(\Omega,\psi) - e_0(\Omega,\psi) = e_0(\Omega+\delta\Omega,\psi+\delta\psi) - e_0(\Omega,\psi) = \frac{\partial e_0}{\partial\Omega}\,\delta\Omega + \frac{\partial e_0}{\partial\psi}\,\delta\psi \quad . \tag{4.61}$$

Since $|v_0|/|v_0'|$ is measurable we find $e_0(\Omega,\psi)$ from the data (4.60) by integration of (4.61).

Experiments should decide upon the practicability of the above method.

## 4.8 Discussion

The analysis reported in this chapter shows that superresolution from optical far-zone data is a physical possibility when the object field arises from a $\lambda$-localized source, i.e., a primary or secondary source that is located in a prescribed region of space with linear dimensions of order one wavelength. A coherent optical system with a photometric scanning device, a data digitizer, and a data-processing facility are the required parts of a superresolving interferometer.

Critical to the assessment of the physical possibility of superresolution are in particular the points that i) a band-limited wave field is at variance with $\lambda$-localized sources, ii) given $\lambda$-localized radiation sources, the spectrum of spatial frequencies of the far field extends to complex-valued frequencies whose real part is in excess of the diffraction limit.

The results presented in Sect.4.4 on the Poynting vector of a band-limited EM field show that plane waves associated with high real-valued planar spatial frequencies are indispensible components of the field of a Gaussian source of lateral dimension $\sigma$ of order one wavelength. The time-averaged energy in free space flows along smooth trajectories whenever $\sigma \gg \lambda$ but meanders when $\sigma \lesssim \lambda$, if the field is composed of all homogeneous-plane-wave-components. If the time-averaged energy flow vector in the EM field of a $\lambda$-localized source is not to deviate from smooth trajectories in free space, the expansion of such a field in terms of only homogeneous plane waves [4.32,63,64] is an approximation valid only when $\sigma \gg \lambda$. Smooth trajectories of the time-averaged energy flux are obtained, however, for *any* value of $\sigma$ by expanding the halfspace field in terms of *complex*-planar-frequency plane waves.

Some of the previous approaches to superresolution and the inverse scattering problem [4.1,16,39,40,41] have approximated the exterior field as a finite linear combination of waves that arise from point sources on a boundary surface or from point sources representing the spatial distribution of the scattering potential inside a continuous scatterer. The results of these approaches are at variance with results [4.31,65,66] that truncated the plane-wave spectrum at the diffraction limit and considered information to be significant to the inverse problem only in as much as it is associated with homogeneous plane waves. However, an *un*band-limited spectrum of real-valued planar spatial frequencies is required for the halfspace representation of the field of a $\lambda$-localized source; the far field of such a source can be represented by plane waves associated with complex-valued spatial frequencies; if the halfspace field is represented by plane waves associated with complex-valued (real-valued) spatial frequencies, any plane wave makes a significant contribution only (not only) where its field is causally related to a $\lambda$-localized source.

The questions that arise from Wolter's research (see Sect.4.1.1) can be answered as follows. There are non-trivial objects given by $\lambda$-localized sources, whose near field can be resolved with superresolving power from far-zone data. There are continuous $\lambda$-localized planar sources that can be reconstructed from far-zone data measured with normal precision, including planar-frequency components of the source the real part of which is in excess of the diffraction limit. The required far-zone data can possibly be measured by an optical interferometer with a shifted reference beam. Using plane waves associated with complex-valued planar spatial frequencies, the point-source field used by Wolter can be extended to include the field of continuous sources localized in a small prescribed region of space.

More research has to be done in particular on the analysis of the information transmitted to the far zone by partial waves associated with complex-valued spatial frequencies. Because these waves belong primarily to a source with small spatial extent, it is more difficult to analyze, in terms of these waves, the field that arises from an extended source, e.g., a planar Gaussian source with a standard deviation $\sigma$ large in comparison to the wavelength. For larger objects and an effective bandlimit of spatial frequencies much below $1/\lambda$ (which is the case, e.g., in electron microscopy) the limit of spatial resolution may be re-examined in view of the point that the sources of scattering may be localized within small prescribed regions of space. Future research may in particular concern I) the expansion of the exterior field in terms of multipole waves with complex-valued degree, and II) the inversion of (4.49) by an analytical rather than an iterative numerical method. An experiment aimed at a specific inverse scattering problem, for example the reconstruction of the distribution of the refractive index in a biological cell from data of optical scattering, could clarify the usefulness of the concept of partial waves associated with complex-valued spatial frequencies.

*Acknowledgments.* I wish to thank Mrs. M.L. Schmidt-Weinmar and Mr. W.B. Ramsay for their contributions to this Chapter, and Dr. R.F. Millar, The University of Alberta, Edmonton, Canada, for his helpful comments on the manuscript. Our research on Superresolution of Electromagnetic Far-Field Data has been supported by the National Research Council of Canada.

## References

4.1 R.F. Millar: SIAM J. Math. Anal. *7*, 131-156 (1976)
4.2 D. Slepian, H.O. Pollak: Bell Syst. Tech. J. *40*, 43-63 (1961)
4.3 J.W. Goodman: "Synthetic-Aperture Optics", in *Progress in Optics*, Vol.VIII, ed. by E. Wolf (North-Holland, Amsterdam, London 1970) pp.1-50
4.4 T.S. Huang, W.F. Schreiber, and O.J. Tretiak: Proc. IEEE *59*, 1589-1609 (1971)
4.5 B.R. Frieden: "Evaluation, Design and Extrapolation Methods for Optical Signals, Based on the Prolate Functions", in *Progress in Optics*, ed. by E. Wolf, Vol.IX (North-Holland, Amsterdam, London 1971) pp.311-407
4.6 C. Pask: J. Opt. Soc. Am. *66*, 68-70 (1976)
4.7 H.A. Ferwerda: Opt. Commun. *3*, 217-219 (1971)
4.8 G.A. Viano: J. Math. Phys. *17*, 1160-1165 (1976)
4.9 H. Wolter: Physica *24*, 457-475 (1958)

4.10 H. Wolter: Opt. Acta *7*, 53-64 (1960)
4.11 H. Wolter: Archiv Elektr. Übertragung *20*, 103-112 (1966)
4.12 B.J. Hoenders, H.A. Ferwerda: Optik *37*, 542-556 (1973)
4.13 H.J. Landau, H.O. Pollak: Bell Syst. Tech. J. *40*, 65-84 (1961)
4.14 H.J. Landau, H.O. Pollak: Bell Syst. Tech. J. *41*, 1295-1336 (1962)
4.15 D. Slepian: Bell Syst. Tech. J. *43*, 3009-3057 (1964)
4.16 G. Euler, S. Blume, H. Wolter: Archiv Elektr. Übertragung *18*, 747-750 (1964)
4.17 H. Pulvermacher: Optik *44*, 413-426 (1976)
4.18 H. Pulvermacher: Optik *45*, 1-10 (1976)
4.19 H. Nassenstein: Opt. Commun. *1*, 146-148 (1969)
4.20 H. Nassenstein: Naturwissenschaften *57*, 468-473 (1970)
4.21 H. Nassenstein: Optik *29*, 597-607 (1969)
4.22 H. Nassenstein: Optik *30*, 44-55 (1969)
4.23 H. Lukosz, A. Wüthrich: Optik *41*, 191-211 (1974)
4.24 H.G. Schmidt-Weinmar: Can. J. Phys. *55*, 1102-1114 (1977)
4.25 H.P. Baltes, H.G. Schmidt-Weinmar: Phys. Lett. *60A*, 275-277 (1977)
4.26 H.G. Schmidt-Weinmar, H.P. Baltes: Helv. Phys. Acta *50*, 669-673 (1977)
4.27 H.G. Schmidt-Weinmar, W.B. Ramsay: Appl. Phys. *14*, 175-181 (1977)
4.28 H.P. Baltes, W.B. Ramsay, H.G. Schmidt-Weinmar: J. Opt. Soc. Am. *67*, 1437 (1977)
4.29 H.G. Schmidt-Weinmar: J. Opt. Soc. Am. *65*, 1059-1066 (1975)
4.30 D. Gabor: "Light and Information", in *Progress in Optics*, ed. by E. Wolf, Vol.I (North-Holland, Amsterdam 1961) pp.109-153
4.31 A.J. Devaney, E. Wolf: J. Math. Phys. *15*, 234-244 (1974)
4.32 W.H. Carter: J. Opt. Soc. Am. *62*, 1195-1201 (1972)
4.33 H. Weyl: Ann. Phys. *59*, 481-500 (1919)
4.34 L.M. Magid: *Electromagnetic Fields, Energy and Waves* (Wiley and Sons, New York 1972)
4.35 A. Sommerfeld: *Partial Differential Equation in Physics* (Academic Press, New York 1964)
4.36 Chen-To Tai: *Dyadic Green's Function in Electromagnetic Theory* (Intext Educational Publishers, Scranton PA, 1971)
4.37 A. Sommerfeld: *Optics* (Academic Press, New York 1964)
4.38 J.W. Goodman: *Introduction to Fourier Optics* (McGraw-Hill, New York 1968)
4.39 H.G. Schmidt-Weinmar: J. Opt. Soc. Am. *61*, 1578 (1971)
4.40 H.G. Schmidt-Weinmar: J. Opt. Soc. Am. *63*, 1307 (1973)
4.41 D.K. Lam, H.G. Schmidt-Weinmar, A. Wouk: Can. J. Phys. *54*, 1925-1936 (1976)
4.42 J.F. Ahner: J. Inst. Math. Its Appl. *19*, 425-439 (1977)
4.43 R.E. Kleinman, W.L. Wendland: J. Math. Anal. Appl. *57*, 170-202 (1977)
4.44 A. Sommerfeld: Ann. Phys. *28*, 665-736 (1909)
4.45 C.J. Bouwkamp, H.B.C. Casimir: Physica *20*, 539-554 (1954)
4.46 A. Baños: *Dipole Radiation in the Presence of a Conducting Half-Space* (Pergamon Press, New York 1966)
4.47 W.W. Hansen: Phys. Rev. *47*, 139-143 (1935)
4.48 C. Müller: *Foundations of Mathematical Theory of Electromagnetic Waves* (Springer, New York, Heidelberg, Berlin 1969)
4.49 J.A. Stratton: *Electromagnetic Theory* (McGraw-Hill, New York 1941)
4.50 G. Goertzel, N. Tralli: *Some Mathematical Methods of Physics* (McGraw-Hill, New York 1960)
4.51 M. Abramowitz, I.A. Stegun: *Handbook of Mathematical Functions* (Dover, New York 1970)
4.52 A. Sommerfeld: *Partielle Differentialgleichungen der Physik* (Geest u. Portig, Leipzig 1948)
4.53 C.H. Wilcox: Comms. Pure Appl. Maths. *9*, 115-134 (1956)
4.54 P.C. Clemmow: *The Plane Wave Spectrum Representation of Electromagnetic Fields* (Pergamon Press, New York 1966)
4.55 F. Noether: "Spreading of Electric Waves Along the Earth", in *Theory of Functions as Applied to Engineering Problems*, ed. by R. Rothe et al. (Massachusetts Inst. of Technology, Cambridge 1951) pp.167-184
4.56 H.G. Schmidt-Weinmar, B. Steinle, H.P. Baltes: "Superresolution of Space-Limited Source Fields by Nonuniform Plane Waves?", to be published in Optik (1978)
4.57 G.C. Sherman, J.J. Stamnes, A.J. Devaney, E. Lalor: Opt. Commun. *8*, 271-274 (1973)

4.58 G.C. Sherman, J.J. Stamnes, E. Lalor: J. Math. Phys. *17*, 760-776 (1976)
4.59 W.T. Wyatt, D.W. Lozier, D.J. Orser: ACM Transaction on Math. Software *2*, 209-231 (1976)
4.60 H.G. Schmidt-Weinmar: J. Opt. Soc. Am. *65*, 999-1002 (1975)
4.61 M. Françon: *Optical Interferometry* (Academic Press, New York 1966)
4.62 H.G. Schmidt-Weinmar: J. Opt. Soc. Am. *63*, 547-555 (1973)
4.63 W.H. Carter: Opt. Commun. *7*, 211-218 (1972)
4.64 W.H. Carter: Opt. Acta *21*, 871-892 (1974)
4.65 E. Wolf: Opt. Commun. *1*, 153-156 (1969)
4.66 A.J. Devaney: J. Opt. Soc. Am. *67*, 1437 (1977)

## Note Added in Proof

Huygens' principle is treated in B.B. Baker, E.T. Copson: *The Mathematical Theory of Huygens' principle* (Clarendon Press, Oxford 1969). To illustrate the general idea, we refer to p.5 of this monograph: "..., the disturbance in the medium which we may for convenience call aether, generally proceeds from sources. These sources are, from the dynamical point of view, singularities at which energy is introduced into the aether; ..."

Our use of the term "precursor" differs from A. Sommerfeld: *Optics* (Academic Press, New York 1972) p.118. Our definition of a precursor is similar to its use when describing the anticipatory response of an ideal linear filter [see, e.g., A.B. Carlson: *Communication Systems*, 2nd ed. (McGraw-Hill, New York 1975) p.68].

The present chapter is concerned with *coherent* subwavelength sources. Superresolution for *partially coherent* light was recently reexamined by PEŘINA et al. [see Ref. 1.33 in Chap.1], who obtained two-point superresolution with two Gaussiam beams in the presence of noise and errors by using a sampling theorem due to FERWERDA and HOENDERS [see Ref.3.80 in Chap.3].

# 5. Radiometry and Coherence

H. P. Baltes, J. Geist, and A. Walther

With 4 Figures

In modern optics, the state of coherence is considered the most relevant characteristic of a radiation field. In the present chapter we study the inverse problem and related questions for scalar planar sources of any state of first- or second-order spatial coherence. We aim at understanding the relationship between source correlations and both classical and modern radiometric properties such as radiant intensity, radiant fluctuation, and first- and second-order degrees of angular coherence. The inverse problem is then to determine correlation between functions in the source plane from such radiometric data.

The classical concept of radiance and the classical Van Cittert-Zernike theorem are known to describe the first-order radiometric properties in the hypothetical limit of zero coherence area. The inverse problem for fully coherent sources has already been developed in Chapters 2-4. The present chapter bridges the gap between these two extreme cases by allowing for partially coherent sources of any degree of first-order coherence and, in addition, presents the basic concepts and relations of second-order radiometry.

We start with a brief historical account followed by a review of the coherence of blackbody radiations, the blackbody serving later as an important example. Next we consider the concepts and relations of *first-order radiometry:* we derive a generalization of the far-zone form of the Van Cittert-Zernike theorem and the definition of generalized radiance. We establish the relations between the source-plane first-order spatial coherence and the first-order radiometric data, namely the radiant intensity and the degree of angular coherence. These relations are illustrated by a variety of model sources. The generalized radiant emittance and the related radiation efficiency are also discussed. We finally turn our attention to *second-order radiometry* which is concerned with second-order correlation, radiant intensity fluctuation, and fluctuation efficiency. Contact with the problem of scattering by random phase screens treated in Chapter 6 is made wherever possible.

Although Section 5.1 was drafted mainly by Geist, Section 5.3 by Walther, and Sections 5.2 and 5.4-6 by Baltes, each author shares the responsibility for the whole chapter.

## 5.1 The Development of Radiometry

This section summarizes the most important events in the development of the field of radiometry. A more detailed presentation of the history of radiometry was published recently [5.1]. Radiometry is understood here as the measurement of any of the quantities that are transferred by optical radiation such as energy, number of photons, momentum, and the ability to produce visual sensations, as well as fluctuations and correlations.

It is convenient to divide radiometry into three periods: classical (scientific), baroque (diverse applications), and modern. The nominal dates dividing these periods are 1900 (Planck's law) and 1960 (the laser). This last date is particularly convenient for three reasons. First, most of the pre-1960 radiometry [5.2] is obsolete due to improvements in electrical and electronic instrumentation. Second, the introduction of the laser marks the beginning of the electro-optics revolution, which is profoundly influencing radiometry as well as other areas of science and technology. Finally, the first contact between radiometry and coherence, namely the study of blackbody coherence, was initiated at this same time.

### 5.1.1 The Classical Period

What we now refer to as radiometry was founded during the period from 1725 to 1760 by Bouguer and Lambert. This work was restricted to the visual effects of visible radiation (light), and comprises the field of photometry [5.3]. Following this period, Scheele, Lambert, Prevost, and others extended radiometry to include heat transfer by light and invisible radiation [5.4]. In this work, thermometers were often used to assess the relative heat content of different radiations in attempts to determine the relation between light and heat. In 1800 Herschel reported the definitive experiment in this regard, a comparison of the relative heat content and visual effects of dispersed solar radiation. His procedure was straightforward. He let solar radiation pass through a slit in a shade and then through a prism onto a table. He compared the temperature rise of liquid in glass thermometers placed at various locations in the dispersed spectrum with the visual effect produced at the same locations. As a result of this investigation he discovered invisible infrared radiation. Similarly, Scheele extended radiometry to the photochemical activity of light (chemical actinometry [5.5]) in connection with investigations of light-induced photodecomposition. In 1801, Ritter reported actinometric measurements using silver nitrate and dispersed solar radiation, and the subsequent discovery of the ultraviolet spectral region.

During the first half of the 19th century, significant progress was made in a number of diverse directions. Radiometry was indirectly extended to include energy transfer by radiation with the demonstration of the equivalence of heat and energy.

The first really sensitive, artificial detectors, radiation thermocouples and thermopiles, were introduced. Bequerel discovered the photovoltaic effect. Gas lighting was becoming a popular artificial illuminant, thereby creating a technical demand for photometric measurements. Therefore attempts were made to develop standards for photometry that were more stable and reproducible than the standard candles in use at the time.

The major emphasis of *source-based* radiometry has been to develop a standard source, which when constructed according to specific formula and operated under specified conditions would provide a pre-determined quantity and quality of light. This goal was advanced considerably during the middle of the 19th century with the development of the concept of a blackbody by Kirchhoff and Stewart. In principle the blackbody would serve as a standard source of radiation when operated at a specified temperature. However, it was not until 1900, after considerable experimental and theoretical activity, that the physical law that describes the quantity and spectral distribution of blackbody radiation was derived by Planck. This of course marked radiometry's greatest contribution to physics, the formation of quantum theory.

Besides Planck's law, the last twenty years of the 19th century also saw Langley's development of the radiation bolometer and atmospheric radiometry. Great improvements in the sensitivity of radiation thermopiles, the first photoelectric cells, Smith's discovery of photoconductivity in selenium, and the development of practical (carbon filament) incandescent lamps, also occurred during this period. Of more importance to radiometry, however, was the development by Ångström in 1893 of the first standard detector of note, an electrically calibrated bolometer. This event marks the beginning of *detector-based* radiometry.

The phenomenological theory of radiative transfer in absorbing and scattering media was developed immediately after the turn of the century [5.6] and is still used in interpreting classical measurements of reflectance and transmittance of diffusing material [5.7-8].

### 5.1.2 The Baroque Period

From the time of Planck until the invention of the laser, radiometry drifted continuously out of the mainstream of physical research into a supporting role in other areas of science and technology [5.9,10]. Key experiments of the classical period were repeated many times with increasing accuracy. Moreover, this period saw the transformation of the blackbody from an object of scientific investigation to the long-sought primary standard source of optical radiation. As the scientific interest in radiometry waned, its technological importance grew. It was employed as a tool in the commercial development of light-related products. Foremost among these were gas-discharge, tungsten-filament, and fluorescent lamps, as well as photoelectric detectors [5.11]. Atmospheric physics and the commercial development of infrared [5.9] and uv-visible [5.12] spectrophotometers for spectrochemical analysis and quality control of visual appearance (colorimetry [5.13]) also benefited from radiometric ideas during this period. Furthermore, a number of independent radiometric measurement systems with special standards and nomenclatures were developed to support these diverse activities. In fact, it is this proliferation of standards and nomenclatures that gives this period its baroque flavor.

Just one example is the use of the word "intensity" in radiometry. This word has been, and is still, used by different groups of physicists concentrating on different aspects of optical radiation, to designate three different concepts, namely power per unit solid angle from a point source, power per unit area crossing a surface, and power per unit projected area per unit solid angle.

Needless to say this overuse of the word "intensity" resulted in much confusion. However, the reaction to this overuse, an attempt to coin entirely new words for radiometric concepts, resulted in even more confusion, with each field of application having its own esoteric nomenclature. For example, "helioscent", "pharosage", "radiosity", "sterance", "pointance", and "areance", are some of the more bizarre examples of words used for the various concepts described above.

Toward the end of this period, problems associated with military applications of infrared radiation [5.14] and the space program were already stimulating renewed interest in the fundamental aspects of radiometry. Of course, the invention of the laser provided an even stronger stimulus, since classical radiometric standards and techniques were not directly applicable to laser power and energy measurements.

### 5.1.3 The Modern Period

The requirements for more convenient and accurate radiometry growing out of the space program resulted in a number of developments in source-based radiometry. Two new primary standard sources of radiation, synchrotron radiation [5.15] and the hydrogen arc [5.16] were developed. Both efforts were motivated by the difficulties encountered in using blackbody radiation as a standard in the vacuum ultraviolet wavelength region.

After a period of relative inactivity, the field of detector-based (electrical calibration) radiometry was developed significantly with stimulation, not only by the space program and laser development, but also by the availability of new instrumentation and construction technology [Ref.5.1, Chap.IV, and references therein]. Considerable progress was also made in unifying the various source-based and detector-based measurement systems [5.17-20]. The long-standing discrepancy [5.21] of greater than one percent between the measured and calculated values of the Stefan-Boltzmann constant was resolved by BLEVIN and BROWN [5.22].

In fact, this last work demonstrates the usually unappreciated importance of coherence in conventional radiometric practice, since diffraction was the second largest source of error in their determination of the Stefan-Boltzmann constant.

It is commonly assumed that blackbody radiation is fully incoherent, but in actual fact it is partially coherent. Moreover, a considerable amount of coherence can be observed in the plane of measurement due to the growth of partial coherence with propagation; and of course, diffraction is a coherence phenomenon.

While classical radiometry is mainly concerned with the average spectral radiant energy density, a new dimension in the field is being opened by the measurement of further statistical properties of the radiation field [5.23-30] such as the average square energy and higher statistical moments, the degree of first-order coherence (amplitude interferometry), and second-order coherence (intensity interferometry). A necessary step in the study of new radiometric properties is the reexamination of generally accepted classical concepts, such as radiance. The reexamination of this particular concept was initiated in 1968 by WALTHER [5.34] in connection with a study of the *radiometry of partially coherent sources* (see Sect.5.3). Theoretical reinvestigations of *blackbody* radiation have been concerned with blackbody coherence (initiated by BOURRET [5.31] in 1960) and size-effects [5.32,33]. The theory of radiative energy transfer in free electromagnetic fields was reinvestigated recently by WOLF [5.35], and Kirchhoff's radiation law under nonequilibrium conditions was also reexamined recently [5.36].

Since the field where radiometry and coherence overlap is quite new, any summary will soon be obsolete. Still, it must be attempted. We observe that the interaction of the two fields has so far caused no major changes in either field. However, radiometrists are beginning to understand that classical radiometry has not dealt with fully incoherent radiation, but rather with partially coherent radiation in situations where the effects of coherence are negligible (it is hoped). With the recent growth of applications involving lasers, fiber optics and integrated optics, it will be necessary to extend radiometric practice to encompass situations where the result depends upon the degree of coherence. Thus we expect so see continued growth in the theoretical and experimental foundations of the classical radiometry of partially coherent sources. Moreover, it is becoming clear that the propagation of certain coherence phenomena can be treated in a radiometric formalism. Therefore, the continued development and future application of these ideas as a supplement to classical radiometric characterizations is a distinct possibility.

## 5.2 Coherence of Blackbody Radiation

Coherence theory and statistical optics on the one hand and size, shape, and proximity effects on the other were the two main incentives for the renewed interest in the venerable field of blackbody radiation. Being the best-known example of a chaotic field (Gaussian process), the free thermal radiation field is reinvestigated in terms of the first-order correlation function [5.29,31,37-49]. The corresponding spatial coherence function serves as an exemplary case in the modern theories of radiometry [5.34] and radiative transfer [5.35]. Moreover thermal radiation is reconsidered as a test case against the validity of semiclassical radiation theory [5.50-54]. Thermal fields in the presence of material walls are studied in connection with the Casimir effect [5.55-57], radiative transfer between closely spaced bodies [5.58,59], eigenvalue spectra of finite cavities [5.29,32,33,60-64], and thermodynamics of the perfect gas in a small enclosure [5.32,33,65-68]. The two aspects of coherence and finite size merge in recent studies of the coherence of bounded thermal radiation fields [5.69-76]. Reviews [5.29,32,33] of the above-mentioned developments are available elsewhere. In the present section we therefore present only a few exemplary results which emphasize coherence properties of the free thermal field and the contact with inverse problems.

### 5.2.1 Temporal Coherence

Blackbody coherence was first investigated by BOURRET [5.31] and KANO and WOLF [5.37] on the basis of Planck's spectral density and techniques, analogous to those employed in the theory of isotropic turbulence in incompressible fluids, and by SARFATT [5.38] who used the density operator method. More comprehensive results were later derived by MEHTA and WOLF [5.39,40,43] and FOX KELLER [5.41].

The first-order correlation function of the electric field is, by definition

$$\Gamma^{(1)}_{\mu\nu}(\underline{r}_1,t_1\underline{r}_2,t_2) = \left\langle E^{(-)}_{\mu}(\underline{r}_1,t_1)E^{(+)}_{\nu}(\underline{r}_2,t_2)\right\rangle \tag{5.1}$$

with the hermitian conjugate field operators $E^{(-)}_{\mu}$, $E^{(+)}_{\nu}$ ($\mu,\nu=1,2,3$) and space and time variables $\underline{r}_1$, $\underline{r}_2$ and $t_1$, $t_2$. The brackets <...> indicate the average with respect to the statistical state of the field. The tensor (5.1) is a function of $\underline{r} \equiv \underline{r}_2 - \underline{r}_1$ and $t \equiv t_2 - t_1$ if the field is homogeneous and stationary. The time correlation is obtained from (5.1) in the special case $\underline{r}_2 = \underline{r}_1$.

The *free thermal radiation field* (temperature T) has the time correlation

$$\Gamma^{(1)}_{\mu\nu}(0,0,0,t) = \delta_{\mu\nu}\,\Gamma(t) \quad , \tag{5.2}$$

where $\delta_{\mu\nu}$ denotes Kronecker's symbol, and where $\Gamma(t)$ is essentially the Fourier transform of Planck's spectrum and obeys

$$\Gamma(t)/\Gamma(0) = \gamma(\tau) = 90\pi^{-4}\sum_{n=1}^{\infty}(n+i\tau)^{-4} \quad . \tag{5.3}$$

We have introduced the reduced time difference $\tau = t(KT/\hbar)$ with K denoting Boltzmann's constant, T the temperature, and $\hbar$ denoting Planck's constant divided by $2\pi$. Function (5.3) is known as the complex degree of temporal coherence of Planck's field.

The modulus $|\gamma(\tau)|$ decreases rapidly (relaxation time of the order of $\hbar/KT$) and shows the asymptotic behavior $|\gamma(\tau)| \sim 30\pi^{-4}\tau^{-3}$ as $\tau \to \infty$. The phase $\Phi(\tau) \equiv \arg\{\gamma(\tau)\}$ has the long-time behavior $\Phi(\tau) \sim -3\pi/2 + (3/2)\tau^{-1}$ [5.37]. We mention that (5.3) is an interesting result with respect to the phase problem in coherence theory (see Sect.2.2.1). The function $\gamma(\tau)$ and hence the energy spectrum of the source is uniquely determined by the modulus $|\gamma(\tau)|$ alone, since $\gamma(z)$, when considered as a function of the complex variable z, has no zeros in the lower half of the complex plane $\mathrm{Im}\, z < 0$ [5.37,77,78].

The *bounded, and hence nonhomogeneous, thermal radiation field* (small cavity, presence of walls) shows a time correlation that depends on the position $\underline{r}_1 = \underline{r}_2$. By virtue of the quantum-optical version of the Wiener-Khintchine theorem [5.79,80], the Fourier conjugate of the spectral energy distribution is then provided by the spatial average

$$(3V)^{-1} \sum_{\mu=1}^{3} \int d^3\underline{r}_1 \; \Gamma_{\mu\mu}^{(1)}(\underline{r}_1, 0, \underline{r}_1, t) \quad , \tag{5.4}$$

where V denotes the cavity volume. This time correlation and the corresponding spectrum were derived by BALTES et al. [5.71] for the thermal radiation field in an empty lossless cavity of cuboidal shape. The underlying mode density (frequency $\omega$) reads

$$\pi^{-2} c^{-3} V\omega^2 \sum_{\nu_1,\nu_2,\nu_3=-\infty}^{+\infty} (2\bar{\nu}\omega/c)^{-1} \sin(2\bar{\nu}\omega/c)$$
$$- (2\pi c)^{-1} \sum_{j=1}^{3} \left[ L_j \sum_{m=-\infty}^{+\infty} \cos(2mL_j\omega/c) \right] + (1/2)\delta(\omega) \quad , \quad \bar{\nu} \equiv \left[ \sum_{j=1}^{3} (\nu_j L_j) \right]^{1/2} \quad , \tag{5.5}$$

where $L_j$ denotes the edge-lengths of the cuboids and c stands for the speed of light. The term with $\bar{\nu} = 0$ corresponds to the free thermal field (thermodynamic limit); all the other terms are corrections of Planck's mode density due to the finite size of the cavity. The harmonic terms ($\bar{\nu} \neq 0, m \neq 0$) in (5.5) describe the "oscillatory" behavior of the density due to the discreteness of the spectrum. The corresponding distribution of the electromagnetic modes in the case of general analytic perfectly conducting boundaries was found only recently by BALIAN and DUPLANTIER [5.64]. Such expressions for the exact mode density offer a key to the solution of the inverse problem of determining the shape of the boundary from the knowledge of the electromagnetic eigenfrequencies [5.62]. Partial solutions of the less involved scalar version of the problem are known [5.32].

### 5.2.2 Spatial Coherence

The complete space-time correlation of the *free* thermal radiation field reads [5.39-42]

$$\Gamma^{(1)}_{\mu\nu}(0,0,\underline{r},t)/\Gamma^{(1)}_{\mu\nu}(0,0,0,0) = 90\pi^{-4}\sum_{n=1}^{\infty}\left\{\delta_{\mu\nu}\left[(n+i\tau)^2+(r/\alpha)^2\right]^{-2}\right.$$
$$\left.+2\alpha^{-2}(r_\mu r_\nu - r^2\delta_{\mu\nu})\left[(n+i\tau)^2+(r/\alpha)^2\right]^{-3}\right\} \tag{5.6}$$

with $\underline{r} = (r_1,r_2,r_3)$, $r = |\underline{r}|$ and $\alpha = \hbar c/KT$. The time correlation (5.3) is recovered from (5.6) by putting $r = 0$. The pure space correlation would be obtained for $\tau = 0$. In view of the quantum-optical Wiener-Khintchine theorem it is, however, more convenient to study spatial coherence in terms of the cross-spectral density or spectral coherence function [5.80], viz.

$$W_{\mu\nu}(\underline{r}_1,\underline{r}_2,\omega) = \int_{-\infty}^{+\infty} dt\, \exp(i\omega t)\Gamma^{(1)}_{\mu\nu}(\underline{r}_1,0,\underline{r}_2,t) \quad . \tag{5.7}$$

A systematic study of the scalar version of this concept was published recently by MANDEL and WOLF [5.81]. The spectral coherence of the (homogeneous) free thermal radiation field is described by [5.43]

$$W_{\mu\nu}(0,\underline{r},\omega) = 2\hbar(\omega/c)^3\left[\exp(\hbar\omega/KT)-1\right]^{-1}\left\{\delta_{\mu\nu}\left[j_0(kr)-(kr)^{-1}j_1(kr)\right]+r^{-2}r_\mu r_\nu j_2(kr)\right\} \quad , \tag{5.8}$$

where $j_s$ denotes the spherical Bessel function of order s and where $k = \omega/c$. Expressing $j_s$ in terms of trigonometric functions, one obtains

$$W(0,\underline{r}) \equiv \sum_{\mu=1}^{\infty} W_{\mu\mu}(0,\underline{r},\omega) = 4\hbar\, k^3\left[\exp(\hbar ck/KT)-1\right]^{-1}(kr)^{-1}\sin kr \tag{5.9}$$

in agreement with SARFATT's result [5.38]. The spectral coherence function (5.9) is used for studying blackbody radiation on the level of scalar physical optics. WALTHER [5.34] observed that Lambert's cosine radiant intensity is connected with spectral coherence of the form (5.9) (see Sect.5.3.4). We note that (5.9) is correct for a large blackbody with sufficiently large aperture, but has to be modified for small apertures where diffraction effects lead to corrections of Lambert's distribution (see Sect.5.4.3).

## 5.3 First-Order Radiometry

Using the scalar approximation for a single mode, we derive the basic laws of radiometry from the theory of partial coherence. In Section 5.3.1 we motivate the need for coherence theory, and derive the relations describing the energy flow in scalar field. Next we develop the radiometric quantities (radiant flux, emittance, and radiance, Sect. 5.3.2) and reexamine the Van Cittert-Zernike theorem (Sect.5.3.3). Examples are discussed in Section 5.3.4.

### 5.3.1 Energy Flow in Scalar Fields

Traditional radiometry, until recently, has been far removed from the mainstream of physical optics. It was based on the assumptions that 1) radiant energy propagates along straight lines, and 2) the energy flow in light fields is an additive quantity. Accepting these suppositions, it follows that the light flux (energy per unit time) moving through an area into a specified solid angle (Fig.5.1) can be calculated by straightforward integration [5.3,82]:

$$\Phi = \iint B \, d\Omega \, dA \cos\vartheta \quad , \tag{5.10}$$

in which the radiance B is the flux per unit solid angle and per unit projected area. B is a function of position as well as direction.

The flux $\Phi$ is expressed in units of power, usually watts. The local flux per unit area, integrated over all directions

$$E = \frac{d\Phi}{dA} = \int B \, d\Omega \, \cos\vartheta \tag{5.10a}$$

is called the emittance if it is radiated by a source surface, and the irradiance if it is received by a surface to be illuminated. The flux per unit solid angle in a specified direction, integrated over the source area,

$$J = \frac{d\Phi}{d\Omega} = \int B \, dA \, \cos\vartheta \tag{5.10b}$$

is called the radiant intensity of the source in that direction.

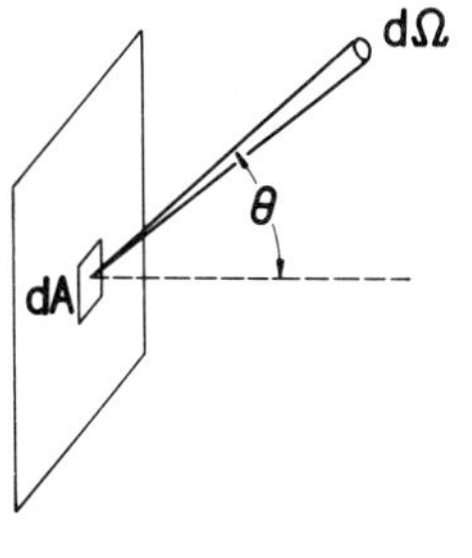


Fig. 5.1. Flux from radiating surface, at obliquity angle $\vartheta$

The additive nature of the energy flux suggests that conventional radiometry can deal with incoherent fields only. It is a well-established fact, however, that incoherent fields do not exist ([5.83], see also [5.84]). Even if they did, incoherent sources could not account for the great variety of radiation patterns observed in nature. The light received in a point at a great distance from a source consists of wavelets produced by all the source points; therefore, the details of a radiation pattern can be explained only by interference, which requires some degree of coherence.

A proper accounting for the interference effects requires the machinery of coherence theory. Restricting ourselves to a scalar theory for one frequency component of a stationary optical field, we describe the first-order coherence properties of the field by its cross-spectral density $W(\underline{r}_1,\underline{r}_2)$, which may be considered as the spatial correlation function of a stochastic amplitude $u(\underline{r})$:

$$W(\underline{r}_1,\underline{r}_2) = \langle u^*(\underline{r}_1)u(\underline{r}_2)\rangle \quad . \tag{5.11}$$

Both W and u depend in general on frequency, but this dependence is not significant for the developments in this section, and will therefore not be displayed in the formulas. The angle brackets indicate an ensemble average that may be calculated, depending on the context, classically, or quantum mechanically. For a formal justification of this rather simplified approach to coherence theory we must refer to the literature, e.g., [5.81], where further references may be found.

The stochastic amplitude $u(\underline{r})$ satisfies Helmholtz's equation, and can therefore be represented by an integral over plane waves

$$u(\underline{r}) = u(x,y,z) = \iint \hat{u}(s_x,s_y)\exp(ik\underline{s}\cdot\underline{r})\,ds_x\,ds_y \quad . \tag{5.12}$$

The vector $\underline{s} = (s_x,s_y,s_z)$ is a unit vector; the sign of its third component will be chosen such that the light propagates in the direction of positive z. The angular spectrum $\hat{u}(s_x,s_y)$ is a function of two variables that, for instance, can be derived from the amplitude distribution in the plane $z = 0$ by Fourier inversion of (5.12) (wavelength $\lambda$):

$$\hat{u}(s_x,s_y) = \frac{1}{\lambda^2}\iint u(x,y,0)\exp[-ik(xs_x+ys_y)]dxdy \quad . \tag{5.13}$$

The quantity of primary interest in radiometry is the energy flow. In a scalar field $V(\underline{r},t)$ that satisfies the wave equation, the energy flow vector must necessarily be of the form [5.85-87]

$$\underline{P} = -\frac{\partial V}{\partial t}\,\mathrm{grad}\,V \quad , \tag{5.14}$$

up to a constant. Specializing to single mode fields

$$V(\underline{r},t) = 2\mathrm{Re}\, u(\underline{r}) \exp(-i\omega t)$$

this reduces to

$$\underline{P} = -i\omega[u^*\mathrm{grad}\, u - u\, \mathrm{grad}\, u^*] \quad . \tag{5.15}$$

To use this result for partially coherent fields, we still have to take an ensemble average. This leads to

$$\underline{P} = 2\omega\ \mathrm{Im}\,\langle u^*\mathrm{grad}\ u\rangle = 2\omega\ \mathrm{Im}\ \mathrm{grad}_1 W(\underline{r},\underline{r}+\underline{r}_1)\Big|_{\underline{r}_1=\underline{0}} \quad . \tag{5.16}$$

In the literature the averaged squared modulus of the amplitude

$$\langle u^*(\underline{r})u(\underline{r})\rangle = W(\underline{r},\underline{r})$$

is often called the intensity. This quantity represents neither an energy flow nor an energy density [5.88]; it is proportional to the response to be expected from a point detector placed at $\underline{r}$, in an observation with a time-bandwidth product greatly in excess of unity. In radiometry this quantity is of little interest; it must not be confused with the radiant intensity defined above. The quantity that controls the energy transfer from one part of space to another is the energy flow vector given by (5.14-16); it is the equivalent of the Poynting vector in the electromagnetic treatment of optical fields.

### 5.3.2 Coherence Theory and the Radiometric Quantities

*Radiant Flux*

We assume that all physical sources are located to the left of the plane $z = 0$. The equations of the previous section can then be used without hesitation in the half space $z > 0$. Our first step towards finding the connections between conventional radiometry and coherence theory consists of calculating the total energy flow in a field described by its cross-spectral density. Far away from the experiment the plane wave integral (5.12) can be approximated by [5.83,89]

$$u(r\underline{s}) = \frac{\lambda}{ir}\, s_z\, \hat{u}(s_x,s_y)\, \exp(ikr) \quad , \quad s_z \equiv \cos\vartheta \quad . \tag{5.17}$$

Substitution into (5.15) yields for the energy flow vector far away from the experiment:

$$\underline{P}(r\underline{s}) = \frac{2\omega k\lambda^2}{r^2} s_z^2 |\hat{u}(s_x,s_y)|^2 \underline{s} \quad . \tag{5.18}$$

A surface element of a sphere centered in the origin and far away from the experiment can be specified by

$$d\underline{A} = \underline{s}\ dA = r^2 \frac{ds_x ds_y}{s_z} \underline{s} \quad . \tag{5.19}$$

Hence the total emitted power is the ensemble average of

$$\Phi = \int \underline{P} \cdot d\underline{A} = 2\omega k\lambda^2 \iint s_z |\hat{u}(s_x,s_y)|^2 ds_x ds_y \quad . \tag{5.20}$$

In radiometry this quantity is referred to as the radiant flux.

*Radiant Intensity*

The radiant intensity of a source in a specified direction is conventionally defined as the flux per unit solid angle. As the quantity $ds_x ds_y/s_z$ is an element of solid angle, we see from (5.20) that the flux emitted per unit solid angle in the direction $\underline{s}$ is given by (up to a constant)

$$J(\underline{s}) = 2\omega k\lambda^2 s_z^2 \left\langle |\hat{u}(s_x,s_y)|^2 \right\rangle \quad ,$$

for which we can write

$$J(\underline{s}) = 2\omega k\lambda^2 s_z^2 \hat{W}(s_x,s_y,s_x,s_y) \quad , \tag{5.21}$$

in which $\hat{W}$ is the full Fourier transform of the cross-spectral density in the plane $z = 0$ (denoted by $W_0$):

$$\hat{W}(\sigma_x,\sigma_y,s_x,s_y) = \frac{1}{\lambda^4} \iint dxdy \iint d\xi d\eta\ W_0(\xi,\eta,x,y) \exp\left[ik(\xi\sigma_x+\eta\sigma_y-xs_x-ys_y)\right] \quad . \tag{5.22}$$

This result was obtained by MARCHAND and WOLF [5.90].

*Radiant Emittance*

The total emitted power can also be calculated by direct integration of the energy flow vector over the plane $z = 0$. This leads to an interesting problem: in order to calculate $\underline{P}(x,y,0)$ in terms of the stochastic field amplitude u, it is necessary to express $\partial u/\partial z$ in terms of $u(x,y,0)$. This is in fact possible, because we have stipulated that the field in the half space $z > 0$ shall propagate to the right. The relation between the field and its gradient is, however, nonlocal. Differentiating (5.12) by z, and substituting (5.13), we find that

$$\frac{\partial}{\partial z} u(x,y,0) = i \iint K(x-\xi,y-\eta)u(\xi,\eta,0)d\xi d\eta \quad , \tag{5.23}$$

in which

$$K(x,y) = \frac{1}{\lambda^2}\iint ks_z \exp[ik(xs_x+ys_y)]ds_x ds_y \quad . \tag{5.24}$$

Restricting the integration to non-evanescent waves, we can evaluate this integral [5.91], with the result that

$$K(x,y) = \frac{2\pi k}{\lambda^2}\sqrt{\frac{\lambda}{2}}\frac{J_{3/2}(k\rho)}{(k\rho)^{3/2}} \quad , \tag{5.25}$$

in which

$$\rho = \sqrt{x^2+y^2} \quad .$$

The z-component of the energy flow vector becomes

$$P_z(x,y,0) = 2\omega \ \mathrm{Re} \ u^*(x,y,0) \iint K(x-\xi,y-\eta)u(\xi,\eta)d\xi d\eta \quad , \tag{5.26}$$

and the total energy flow per second is, after once again taking the ensemble average:

$$\Phi = 2\omega \iint dxdy \iint d\xi d\eta \ \mathrm{Re} \ W_0(x,y,\xi,\eta)K(x-\xi,y-\eta) \quad . \tag{5.27}$$

Equations (5.20) and (5.27) are Fourier conjugates.

At this point one might, in analogy to the step leading from (5.20) to (5.21), wish to ask for the amount of energy radiated per unit area of the source surface $z = 0$. This quantity is conventionally known as the radiant emittance E. It is simply the ensemble average of the z-component of the energy flow vector:

$$E(x,y,0) = P_z(x,y,0) = 2\omega \iint d\xi d\eta \ \mathrm{Re} \ W_0(x,y,\xi,\eta)K(x-\xi,y-\eta) \quad . \tag{5.28}$$

However, the nonlocal nature of this quantity leads to the somewhat disconcerting result that when a diaphragm is placed over the plane $z = 0$, the radiant emittance is not only reduced to zero on the opaque part of the diaphragm, but is also changed numerically in the transmitting region. Quantitative examples of such pecularities are treated in Section 5.5. Equation (5.28) is similar in essence to a result obtained by MARCHAND and WOLF [5.90], but is different in detail, because their derivation was based on an early definition of radiance, stated in the next section as (5.34). MARCHAND and WOLF show also in [5.90] that the emittance, on occasion, can be negative in some spots of the source plane, a subject further discussed in [5.92,93].

*Radiance*

The radiance is conventionally defined as the flux per unit solid angle and per unit projected area, i.e., the intensity per unit area, or the emittance per unit solid angle. The difficulties now compound themselves, because subdividing the emittance into contributions from various directions is not a very well-defined process. To solve the problem WALTHER [5.94] considered the flux passing through any element of surface $\underline{n}\, dA$ located in the half space $z > 0$. This flux is given by

$$d\Phi = \left\langle \underline{P} \cdot \underline{n}\, dA \right\rangle = 2\omega \, \mathrm{Im} \left\langle u^* \mathrm{grad}\, u \right\rangle \cdot \underline{n}\, dA \quad . \tag{5.29}$$

Representing u by an integral over plane waves yields

$$d\Phi = 2\omega\, k\, dA \iint \underline{s} \cdot \underline{n}\, \mathrm{Re} \left\langle u^*(\underline{r}) \hat{u}(s_x, s_y) \right\rangle \exp(ik\underline{s} \cdot \underline{r}) ds_x ds_y \quad . \tag{5.30}$$

We recognize $\underline{s} \cdot \underline{n}$ as the cosine factor occurring in (5.10). Also, $ds_x ds_y / s_z$ is an element of solid angle. So we can write the last equation in the form

$$d\Phi = dA \int 2\omega\, k s_z\, \mathrm{Re} \left\langle u^*(\underline{r}) \hat{u}(s_x, s_y) \right\rangle \exp(ik\underline{s} \cdot \underline{r}) d\Omega \cos\vartheta \quad . \tag{5.31}$$

This suggests that we might take for the physical optics definition of the radiance

$$B(\underline{r}, \underline{s}) = 2\omega\, k s_z\, \mathrm{Re} \left\langle u^*(\underline{r}) \hat{u}(s_x, s_y) \right\rangle \exp(ik\underline{s} \cdot \underline{r}) \quad . \tag{5.32}$$

It can be shown [5.94] that this definition of the radiance is invariant under coordinate rotations.

If we restrict ourselves to the plane $z = 0$, (5.32) can be written as

$$B(x,y,s_x,s_y) = \frac{k\omega}{\lambda^2} s_z \iint \left[ W_0(x,y,x+\xi,y+\eta) + W_0(x-\xi,y-\eta,x,y) \right] \cdot \exp[-ik(\xi s_x + \eta s_y)] d\xi d\eta \quad . \tag{5.33}$$

It follows from their very definition that the emittance E as given by (5.28) and the radiance B as defined by (5.32) are interrelated by the conventional relations that were stated in the opening paragraphs of Section 5.3.1. For the electromagnetic field, quantities similar to those defined above were introduced by WOLF [5.35].

In an earlier paper WALTHER [5.34] used a slightly different definition for the radiance:

$$B^{(1)}(x,y,s_x,s_y) = \frac{k\omega}{\lambda^2} s_z \iint 2W\left(x - \frac{1}{2}\xi, y - \frac{1}{2}\eta, x + \frac{1}{2}\xi, y + \frac{1}{2}\eta\right) \cdot \exp[-ik(\xi s_x + \eta s_y)] d\xi d\eta \quad . \tag{5.34}$$

This definition looks attractive because it defines a Wigner distribution [5.95] on the configuration space $(x,y,s_x,s_y)$. It can be proved that an integral of the type (5.10) over the plane $z=0$, using $B^{(1)}$ rather than B, does in fact yield the correct radiant flux. However, a decisive disadvantage of $B^{(1)}$ is that its definition is not invariant under coordinate rotations around the x- and y-axis [5.92-94]. Moreover, $B^{(1)}$, on occasion, can be nonzero in opaque points of a diaphragm placed over a source; B as defined by (5.32) does not suffer from this peculiarity.

A very important property of the conventional radiance is its constancy along light rays. Using the plane wave expansion for $u^*(\underline{r})$, (5.32) can be cast in the form

$$B(\underline{r},\underline{s}) = \frac{2\omega k}{\lambda^2} s_z \operatorname{Re} \iint \hat{W}(\sigma_x,\sigma_y,s_x,s_y)\exp[ik(\underline{s}-\underline{\sigma})\cdot\underline{r}]d\sigma_x d\sigma_y \quad , \tag{5.35}$$

in which $\underline{\sigma}$ is again a unit vector. We now calculate the value of B in points of a straight line passing through $\underline{r}$ in the direction $\underline{s}$, i.e., in the points $\underline{r}+p\underline{s}$. The parameter p measures the distance from $\underline{r}$ to the new point on the line. Substitution yields

$$B(\underline{r}+p\underline{s},\underline{s}) = \frac{2\omega k}{\lambda^2} s_z \operatorname{Re} \iint \hat{W}(\sigma_x,\sigma_y,s_x,s_y)\exp[ikp(1-\underline{\sigma}\cdot\underline{s})]\exp[ik(\underline{s}-\underline{\sigma})\cdot\underline{r}]d\sigma_x d\sigma_y . \tag{5.36}$$

If the angular correlation function is substantially zero unless $\underline{s}$ and $\underline{\sigma}$ have very nearly the same direction, the exponential containing p may be replaced by unity for not too large values of p. Under these conditions (5.36) reduces to (5.35), so that B becomes invariant along straight lines, that may well be called light rays.

### 5.3.3 The Van Cittert-Zernike Theorem

In the previous section we derived expressions to calculate the radiometric quantities $\Phi$, J, E, and B for fields specified by their cross-spectral density. The problem was phrased the other way around in the time before modern coherence theory attained its paramount position in optics: given a distribution of supposedly incoherent sources in space, what are the coherence properties of the field created? This problem was solved by VAN CITTERT and ZERNIKE in papers that are now classics [5.96-98]. VAN CITTERT studied in detail the joint probability distribution for the phase in pairs of points in the field. ZERNIKE avoided this detailed treatment by primarily directing his attention to the prediction of the results of interference experiments. Mathematically, their results were identical. In VAN CITTERT's words [Ref.5.99, Sect.28, pp.440,442]:

> "(1) The degree of coherence in a plane illuminated by a light-source is identical with the diffraction function of a diaphragm similar in form and size to the light source.
>
> (2) If the vibrations at two points in a plane, illuminated by a light source, are made to interfere, an interference system is obtained of which the visibility is equal to the degree of coherence between the points."

These rules, known as the Van Cittert-Zernike theorem, acquire a new significance in the light of the previous section. Central to this development is the physical meaning of the Fourier transform of the cross-spectral density introduced in the previous section by (5.22). Starting with the far-field representation (5.17) for the stochastic amplitude, we can write for the cross-spectral density in the far field:

$$\langle u^*(r_1\underline{s}_1)u(r_2\underline{s}_2)\rangle = \lambda^2 \frac{\exp ik(r_2-r_1)}{r_2r_1} s_{z1}s_{z2}\,\hat{W}(s_{x1},s_{y1},s_{x2},s_{y2}) \quad . \tag{5.37}$$

The dependence of this correlation function on $r_1$ and $r_2$ is clearly trivial. The physically significant aspect of the formula is that $\hat{W}(s_{x1},s_{y1},s_{x2},s_{y2})$ represents the correlation between the amplitudes in points on a sphere far away from the experiment, i.e., the angular correlation of the field. We now turn to (5.22), and introduce the new variables

$$\begin{aligned} x_1 &= x+\xi \;, & y_1 &= y+\eta \;, & s_{x1} &= s_x \;, & s_{y1} &= s_y \;, \\ x_2 &= x \;, & y_2 &= y \;, & s_{x2} &= s_x+s_x' \;, & s_{y2} &= s_y+s_y' \;. \end{aligned}$$

Substitution in (5.22) yields

$$\hat{W}(s_x,s_y,s_x+s_x',s_y+s_y') = \frac{1}{\lambda^4}\iint W_0(x+\xi,y+\eta,x,y)\exp[ik(\xi s_x+\eta s_y-xs_x'-ys_y')]dxdyd\xi d\eta . \tag{5.38}$$

This relation is once again a Fourier transform. We see that the following pairs of variables are Fourier conjugates:

$$\begin{aligned} &s_x \text{ and } \xi \;, && s_x' \text{ and } x \;, \\ &s_y \text{ and } \eta \;, && s_y' \text{ and } y \;. \end{aligned}$$

In a broad sense, the dependence of $W_0$ on the difference variables $\xi$ and $\eta$ represents the coherence properties of the source plane $z = 0$. The dependence of $W_0$ on the position variables x and y represents, in essence, the intensity distribution over the source. Similarly, the dependence of $\hat{W}$ on the difference variables $s_x'$ and $s_y'$ represents the angular coherence in the far field, while the dependence of $\hat{W}$ on $s_x$ and $s_y$ represents the intensity distribution in the far field, i.e., the radiant intensity distribution. The Fourier relation (5.38), therefore, implies that the *radiant intensity* of a source is Fourier conjugate to its *coherence properties*, while the *angular coherence* (the coherence properties in the far field) is Fourier conjugate to the *intensity distribution* over the source. This is in substantial agreement with the Van Cittert-Zernike theorem. In specific cases the general statements made in this paragraph can, of course, be made much more specific. We refer in particular to Sections 5.4 and 5.5, where numerous references to the literature are given.

### 5.3.4 An Example: Quasistationary Sources

For many sources the cross-spectral density $W_0(x,y,x+\xi,y+\eta)$ is substantially zero unless $\xi$ and $\eta$ stay within a small coherence area, and is furthermore independent of x and y for all points of the source not too close to the edge. When these conditions are satisfied, we write $W_0(\xi,\eta)$ for the cross-spectral density. Equation (5.33) then reduces to

$$B(x,y,s_x,s_y) = \frac{2\omega k}{\lambda^2} s_z \iint W_0(\xi,\eta)\exp[-ik(\xi s_x+\eta s_y)]d\xi d\eta \quad , \tag{5.39}$$

i.e., the radiance is proportional to the Fourier transform of the cross-spectral density function, multiplied by the cosine factor $s_z$. We show three examples: I) the limiting case of a fully incoherent source, II) a blackbody source, and III) a random phase screen illuminated with a coherent plane wave.

I) *Incoherent Sources*

By an incoherent source we mean a source with a coherence range much smaller than the wavelength of the light. It is then convenient to write as an approximation

$$W_0(\xi,\eta) = w_0\delta(\xi)\delta(\eta) \quad . \tag{5.40}$$

Equation (5.39) yields

$$B = \frac{2\omega k}{\lambda^2} w_0 s_z \quad . \tag{5.41}$$

Hence, the radiance of an incoherent source tapers off with the cosine of the angle from the normal to the source plane. It follows that the radiant intensity of such a source varies with the square of the cosine of the same angle, a result due to BERAN and PARRENT [5.83].

II) *Blackbody Radiation*

In this case we have, according to (5.9),

$$W_0(\xi,\eta) = w_1 \frac{\sin k\rho}{k\rho} \quad , \tag{5.42}$$

with

$$\rho = \sqrt{\xi^2+\eta^2} \quad .$$

Substitution in (5.39) yields, after transformation to polar coordinates and integration over the polar angle,

$$B = \frac{4\pi\omega k}{\lambda^2} w_1 s_z \int_0^\infty \frac{\sin k\rho}{k\rho} J_0\left(k\rho\left[s_x^2+s_y^2\right]^{1/2}\right)\rho d\rho \tag{5.43}$$

This integral can be evaluated [Ref.5.91, p.487, Eq.(11.4.38)], with the remarkable result that the entire expression reduces to a constant:

$$B = \frac{2\omega}{\lambda} w_1 \quad . \tag{5.44}$$

Hence, the radiance of a blackbody radiator is independent of direction. As a consequence, the radiant intensity of a planar blackbody source varies with a cosine factor to the first power. Sources with this property are known as Lambertian. A more accurate treatment for small-size sources may be found in Section 5.4.

III) *The Random Phase Screen*

As a third example, we consider a random phase screen in the plane $z=0$, with the amplitude transmission

$$T(x,y) = \exp i\varphi(x,y) \quad , \tag{5.45}$$

in which $\varphi(x,y)$ is an isotropic zero mean Gaussian random function with a correlation function $R(\rho)$, $\rho = (\xi^2+\eta^2)^{1/2}$. If the screen is illuminated by the plane wave

$$u_i(x,y,z) = u_0 \exp(ik\underline{\sigma}\cdot\underline{r})$$

the amplitude distribution in the plane $z=0$ beyond the phase screen is

$$u(x,y,0) = u_0 \exp[ik(x\sigma_x+y\sigma_y)]\exp[i\varphi(x,y)] \quad ,$$

so that

$$u^*(x,y,0)u(x+\xi,y+\eta,0) = u_0^2 \exp[ik(\xi\sigma_x+\eta\sigma_y)]\exp(i[\varphi(x+\xi,y+\eta)-\varphi(x,y)]) \quad . \tag{5.46}$$

The average of the results of a large number of identical radiometric measurements carried out with different phase screens of the same type can be derived by considering the ensemble average of the last equation. The average of the left hand side takes on the character of a cross-spectral density. The ensemble average of the last exponential is [5.100]

$$\langle\exp(i[\varphi(x+\xi,y+\eta)-\varphi(x,y)])\rangle = \exp(-[R(0)-R(\rho)]) \quad .$$

We can therefore write

$$W_0(\xi,\eta) = u_0^2 \exp[ik(\xi\sigma_x+\eta\sigma_y)]\exp(-[R(0)-R(\rho)]) \quad . \tag{5.47}$$

Substitution in (5.39) yields

$$B(x,y,s_x,s_y) = \frac{2\omega k}{\lambda^2} u_0^2 s_z \iint \exp(-[R(0)-R(\rho)])\exp(ik[\xi(s_x-\sigma_x)+\eta(s_y-\sigma_y)])d\xi d\eta . \tag{5.48}$$

We see then that the radiance beyond the phase screen is primarily determined by the Fourier transform of the amplitude transmission correlation function, centered on the direction of the incident light. Note, however, that this radiance is an ensemble average; in the individual experiments there will be considerable speckle (see Chap.6). In practical cases (5.48) should be used with caution. It is based on the assumption that the profile of the ground glass is simply impressed on the transmitted wave front, an approximation that can be grossly in error for highly irregular surfaces.

## 5.4 Radiant Intensity and Angular Coherence

In this section we continue the study of first-order spatial coherence of radiation fields by planar sources. We emphasize the evaluation of the radiant intensity and the degree of angular coherence in terms of inverse relations (Sect.5.4.2) and explicit source models with "adjustable" degree of coherence (Sect.5.4.1,3,4). An application to thermionic emission sources is also presented (Sect.5.4.5).

The classic text for partial coherence is found in books by BORN and WOLF [5.101], BERAN and PARRENT [5.83], and O'NEILL [5.102]. The important case of partially coherent aperture field sources is treated in the work of SCHELL [5.103], SHORE et al. [5.104,105], and JAISWAL et al. [5.106,107]. WALTHER's studies [5.34,93,94,108] on coherence and radiometry and the quest for the reconstruction of the source correlation from far-zone data stimulated the current work by WOLF [5.90,92,109-115], and BALTES [Ref.5.29, Chap.3, 5.116-123] and their coworkers. The optical studies of the relation between source-plane and far-zone coherence functions are complemented by current investigations of partial coherence in electron microscopy due to FERWERDA and VAN HEEL [5.124-126], see also [5.127]. The underlying mathematical methods are closely related to those of speckle and scattering theory [5.28,128-134]. Recent measurements of spatial coherence are reported in the literature [5.135-142]. Also of current interest is the related problem of partially coherent imaging [5.143-146].

As in the preceding section, we work with a single Fourier component (frequency $\omega$, wavenumber $k$) and describe the field by the scalar spectral coherence function (5.11). Related vector field calculations were recently presented by CARPENTER and

PASK [5.147,148] and LEADER [5.149,150]. We introduce a slight modification of the notation: from now on, $\underline{s}$ denotes the 2D-vector $(s_x,s_y)$ rather than the 3D-vector of Section 5.3.

### 5.4.1 Source Models

In the source plane $z=0$ we specify the coherence function by

$$W_0(\underline{\varrho}_1,\underline{\varrho}_2) \equiv W(\underline{r}_1,\underline{r}_2)\Big|_{z_1=z_2=0} \tag{5.49}$$

with $\underline{\varrho}_j=(x_j,y_j)$. Assuming a stationary stochastic process and normalizing as usual, the source-plane coherence function can be written as

$$W_0(\underline{\varrho}_1,\underline{\varrho}_2) = \left[I_0(\underline{\varrho}_1)I_0(\underline{\varrho}_2)\right]^{1/2}\mu_0(\underline{\varrho}) \quad , \tag{5.50}$$

$$\underline{\varrho} \equiv \underline{\varrho}_1 - \underline{\varrho}_2 = (x_1-x_2,y_1-y_2) \quad , \tag{5.51}$$

$$I_0(\underline{\varrho}_1) \equiv W_0(\underline{\varrho}_1,\underline{\varrho}_1) \quad , \tag{5.52}$$

in agreement with SCHELL's theorem [5.103, see also 5.104,105,107,114-123,128-134]. Here, $I_0(\underline{\varrho}_1)$ describes the intensity profile in the source plane, and $\mu_0(\underline{\varrho})$ denotes the degree of coherence obeying $\mu_0(0)=1$. Let us call the planar source described by (5.50) a *Schell's source*. This source represents an aperture field source of any degree of coherence, and is the natural generalization of the coherent aperture field source studied in Chapter 4. Schell's source is usually assumed to be bounded with respect to both the intensity distribution and the degree of coherence. Such a source can be characterized by its beam cross section and coherence area. Radiometric properties of Schell's sources were recently studied by JAISWAL et al. [5.107] and BALTES et al. [5.119-123]. Indirect Schell's sources can be produced by illuminating a quasi-planar fluctuating scatterer by a coherent beam. Laser illumination often implies a Gaussian beam profile (width a), viz.,

$$I_0(\underline{\varrho}_1) \propto \exp(-|\underline{\varrho}_1|^2/2a^2) \quad . \tag{5.53}$$

If the intensity profile does not vary appreciably over distances where the degree of coherence differs sensibly from zero, (5.50) is approximated by [5.114,115,121, 126,127]

$$W_0(\underline{\varrho}_1,\underline{\varrho}_2) = I_0(\underline{\varrho}')\mu_0(\underline{\varrho}) \quad , \tag{5.54}$$

$$\underline{\varrho}' = (\underline{\varrho}_1+\underline{\varrho}_2)/2 \quad . \tag{5.55}$$

The above *quasistationary source* leads to simple radiometric relations, as was first noticed by WALTHER [5.34] and later developed by WADAKA and SATO [5.143], BALTES and STEINLE [5.120], FERWERDA and VAN HEEL [5.125,126], and, most comprehensively, by CARTER and WOLF [5.114,115]. The approximation (5.54) is valid for e.g., the blackbody source with aperture of radius $a \gg k^{-1}$ [5.119], or for the virtual electron source [5.125,126]. The same approximation can, however, not account for illuminated random-phase screens with coherence area and beam cross section showing the same order of magnitude as, for example, in the experiments by BERTOLOTTI et al. [5.139-141]. A systematic comparison of the Schell's source (5.50) and the quasistationary source (5.54) was published by BALTES and STEINLE [5.121].

Two more specific families of (isotropic) sources are currently being studied, namely the *Gauss-correlated sources* [5.29,111,114,115,120-123] showing the degree of coherence

$$\mu_0(\underline{\rho}) = \exp(-\rho^2/2b^2) \quad , \quad \rho \equiv |\underline{\rho}| \quad , \quad b > 0 \quad , \tag{5.56}$$

with coherence length b, and the *Bessel-correlated sources* [5.29,117-123] where

$$\mu_0(\underline{\rho}) = (n/2)!(k\rho/2)^{-n/2} J_{n/2}(k\rho) \quad , \quad n > -1 \quad , \tag{5.57}$$

with $J_{n/2}$ denoting the Bessel function of order n/2. The Gaussian degree of coherence is useful for describing the deep random phase screen [5.133,134] (see Sect. 6.3.1). The quest for an expansion of the radiant intensity into a series of cosines [5.116] leads to the Bessel-type degree of coherence, which includes the blackbody [5.34,112] in the case $n = 1$. The coherence area of the Gauss-correlated source is proportional to $b^2$, while the coherence area of the Bessel-correlated source is approximately proportional to n for $n \gtrsim 2$ [5.116]. For small $b^2$, the Gaussian degree of coherence is a candidate for the quasistationary approximation (5.54), but includes a large contribution from high spatial frequencies [5.117,121-123]. Bessel-correlated sources show the complementary behavior; their degree of coherence includes no spatial frequencies above k, and decays slowly with increasing distance $\underline{\rho}$. Further details are presented in Sections 5.4.3 and 5.4.4.

### 5.4.2 Inverse Relations

The art of reconstructing a source coherence function from radiometric data is much less advanced than the state of the corresponding inverse problems of wave optics as presented in Chapters 3 and 4. Neither the coherence-function counterpart of inverse scattering (3D reconstruction) nor that of extrapolation and related expansions has been studied hitherto. The reconstruction of 2D coherence functions is currently investigated [Ref.5.29, Chap.3; 5.112,120], but only methods analogous to "Fourier optics" or "inverse diffraction" [5.151] have been developed hitherto. Moreover, far-zone data from planar sources are evaluated [5.128-131] in terms of specific source models, i.e., taking enormous prior knowledge for granted.

Let us express far-zone positions $\underline{r}$ by spherical coordinates, viz.,

$$\underline{r} = r(\sin\vartheta\cos\varphi, \sin\vartheta\sin\varphi, \cos\vartheta) = r\left[s_x, s_y, (1-s_x^2-s_y^2)^{1/2}\right] = r\left[\underline{s}, (1-s^2)^{1/2}\right] \quad (5.58)$$

with $r = |\underline{r}|$ and with the direction specified by

$$\underline{s} = (s_x, s_y) = \sin\vartheta(\cos\varphi, \sin\varphi) \quad , \quad s = |\underline{s}| = \sin\vartheta \quad . \quad (5.59)$$

We denote the far-zone coherence function by

$$W_{r_1r_2}(\underline{s}_1, \underline{s}_2) = W(\underline{r}_1, \underline{r}_2)\Big|_{\underline{r}_j = r_j\left[\underline{s}_j, (1-s_j^2)^{1/2}\right]} \quad . \quad (5.60)$$

Using Green's functions [5.23,83,102] or the angular spectrum representation [5.94, 106-109], one can derive the (approximative) relation

$$W_{r_1,r_2}(\underline{s}_1, \underline{s}_2) = (k/2\pi)^2 (r_1 r_2)^{-1} \exp[ik(r_1-r_2)] \cos\vartheta_1 \cos\vartheta_2$$
$$\int d^2\underline{\rho}_1 \int d^2\underline{\rho}_2 \exp\left[-ik(\underline{s}_1\underline{\rho}_1 - \underline{s}_2\underline{\rho}_2)\right] W_0(\underline{\rho}_1, \underline{\rho}_2) \quad (5.61)$$

between far-zone and source-plane coherence functions (see Fig.5.2). The low-spatial-frequency part $W_0^{LF}(\underline{\rho}_1, \underline{\rho}_2)$ of the coherence function in the source plane is recovered from a given far-zone function $W_{r_1r_2}(\underline{s}_1, \underline{s}_2)$ by virtue of the *inverse relation* [5.120]

$$W_0^{LF}(\underline{\rho}_1, \underline{\rho}_2) \propto r_1 r_2 \exp\left[-ik(r_1-r_2)\right] \int d^2\underline{s}_1 \exp(ik\underline{\rho}_1\underline{s}_1) \mathrm{circ}(\underline{s}_1)(1-\underline{s}_1^2)^{-1/2}$$
$$\int d^2\underline{s}_2 \exp(ik\underline{\rho}_2\underline{s}_2) \mathrm{circ}(\underline{s}_2)(1-\underline{s}_2^2)^{-1/2} W_{r_1r_2}(\underline{s}_1, \underline{s}_2) \quad , \quad (5.62)$$

where the circle function provides the cut-off $s_j = \sin\vartheta_j \leqq 1$. The remaining high-frequency part

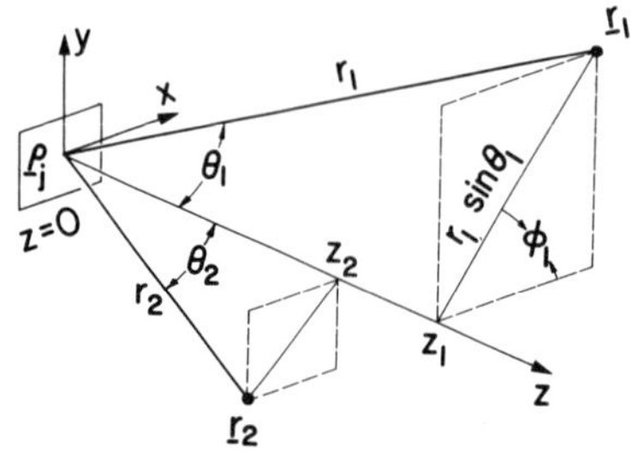


Fig. 5.2. Illustrating the notation

$$W_0^{HF}(\underline{\rho}_1,\underline{\rho}_2) = W_0(\underline{\rho}_1,\underline{\rho}_2) - W_0^{LF}(\underline{\rho}_1,\underline{\rho}_2) \tag{5.63}$$

is responsible for the "fine structure" of the source coherence function and corresponds to complex angles $\vartheta_j$ with $s_j = \sin\vartheta_j > 1$. The problem of recovering information on the function (5.63) is not further pursued in this chapter (see [5.128]).

Since the absolute distances $r_1$ and $r_2$ occur in (5.62) in a trivial way, we choose $r_1 = r_2 = r$ and shall henceforth consider the lateral coherence function

$$W_r(\underline{s}_1,\underline{s}_2) \equiv W_{rr}(\underline{s}_1,\underline{s}_2) \quad . \tag{5.64}$$

Introducing the relative coordinates (5.51,55) and

$$\underline{s}' = \underline{s}_1 - \underline{s}_2 \quad , \quad \underline{s} = (\underline{s}_1+\underline{s}_2)/2 \tag{5.65}$$

we obtain

$$W_r(\underline{s}_1,\underline{s}_2) = (k/2\pi)^2 r^{-2} \cos\vartheta_1 \cos\vartheta_2 \, F(\underline{s}',\underline{s}) \tag{5.66}$$

with the *angular correlation* [cf.(5.22)]

$$F(\underline{s}',\underline{s}) \equiv \int d^2\underline{\rho}' \exp(-ik\underline{s}'\underline{\rho}') \int d^2\underline{\rho} \exp(-ik\underline{s}\underline{\rho}) W_0(\underline{\rho}'+\underline{\rho}/2,\underline{\rho}'-\underline{\rho}/2) \tag{5.67}$$

and the *degree of angular coherence*

$$\begin{aligned} \mu_r(\underline{s}_1,\underline{s}_2) &\equiv W_r(\underline{s}_1,\underline{s}_2)\left[W_r(\underline{s}_1,\underline{s}_1)W_r(\underline{s}_2,\underline{s}_2)\right]^{-1/2} \\ &= F(\underline{s}',\underline{s})\left[F(0,\underline{s}_1)F(0,\underline{s}_2)\right]^{-1/2} \quad . \end{aligned} \tag{5.68}$$

For $\underline{s}_1 = \underline{s} = \underline{s}_2$, (5.66) yields the *radiant intensity* [cf.(5.21)]

$$J(\underline{s}) \equiv r^2 W_r(\underline{s},\underline{s}) = (k/2\pi)^2 \cos^2\vartheta \, F(\underline{s}) \tag{5.69}$$

in terms of the *angular frequency spectrum*

$$F(\underline{s}) \equiv F(0,\underline{s}) = \int d^2\underline{\rho}' \int d^2\underline{\rho} \exp(-ik\underline{s}\underline{\rho}) W_0(\underline{\rho}'+\underline{\rho}/2,\underline{\rho}'-\underline{\rho}/2) \quad . \tag{5.70}$$

The functions (5.67) and (5.70) are crucial for the relationship between far-zone and source-plane properties. The inverse relation (5.62) can be written as [5.120]

$$W_0^{LF}(\underline{\rho}'+\underline{\rho}/2,\underline{\rho}'-\underline{\rho}/2) \propto \int d^2\underline{s}' \exp(ik\underline{\rho}'\underline{s}') \int d^2\underline{s} \exp(ik\underline{\rho}\underline{s})$$

$$\mathrm{circ}(\underline{s}+\underline{s}'/2)\mathrm{circ}(\underline{s}-\underline{s}'/2)\mu_r(\underline{s}+\underline{s}'/2,\underline{s}-\underline{s}'/2)$$

$$\left[J(\underline{s}+\underline{s}'/2)\right]^{1/2}\left[J(\underline{s}-\underline{s}'/2)\right]^{1/2}\left[1-(\underline{s}+\underline{s}'/2)^2\right]^{-1/2}\left[1-(\underline{s}-\underline{s}'/2)^2\right]^{-1/2} \quad . \tag{5.71}$$

Thus knowledge of both the degree of angular coherence $\mu_r(\underline{s}_1,\underline{s}_2)$ and the radiant intensity $J(\underline{s})$ is, in general, required for the retrieval of the source-plane coherence function $W_0^{LF}$. As for the phase of $\mu_r$, we refer to Chapter 2.

The only prior knowledge used this far is the planar geometry of the source. Let us now require more prior knowledge, i.e., consider more specific source models. We start with the fairly general *Schell's source* (5.50) showing [5.107,119-122] the angular correlation

$$F(\underline{s}',\underline{s}) = \int d^2\underline{\rho} \exp(-ik\underline{s}\underline{\rho})\mu_0(\underline{\rho}) \int d^2\underline{\rho}' \exp(-ik\underline{s}'\underline{\rho}')\left[I_0(\underline{\rho}'+\underline{\rho}/2)I_0(\underline{\rho}'-\underline{\rho}/2)\right]^{1/2} \tag{5.72}$$

and the angular spectrum

$$F(\underline{s}) = \int d^2\underline{\rho} \exp(-ik\underline{s}\underline{\rho})\mu_0(\underline{\rho}) \int d^2\underline{\rho}'\left[I_0(\underline{\rho}'+\underline{\rho}/2)I_0(\underline{\rho}'-\underline{\rho}/2)\right]^{1/2}$$
$$= \mathcal{F}\{\mu_0\} \otimes \left[\mathcal{F}\left\{I_0^{1/2}\right\}\right]^2 \tag{5.73}$$

with $\mathcal{F}$ indicating the 2D Fourier transform and $\otimes$ the convolution. The crucial deconvolution of these integrals has been achieved numerically for the case of Bessel-correlated sources with circular aperture [5.119] and analytically for Gaussian-correlated sources with Gaussian beam profile [5.120-123], see Sections 5.4.3 and 5.4.4.

In the special case of the *quasistationary source* (5.54) the less involved expression [5.34,114,120,125]

$$F(\underline{s}',\underline{s}) = \int d^2\underline{\rho}' \exp(-ik\underline{s}'\underline{\rho}')I_0(\underline{\rho}') \int d^2\underline{\rho} \exp(-ik\underline{s}\underline{\rho})\mu_0(\underline{\rho}) \tag{5.74}$$

is valid (see also Sect.5.3). The corresponding angular spectrum reads

$$F(\underline{s}) = S\int d^2\underline{\rho} \exp(-ik\underline{s}\underline{\rho})\mu_0(\underline{\rho}) \quad , \tag{5.75}$$

where

$$S \equiv \int d^2\underline{\rho}' I_0(\underline{\rho}') \tag{5.76}$$

denotes the effective surface area. From (5.69,75) we easily obtain the *inversion of the radiant intensity* [Ref.5.29, Chap.3;5.112]

$$\mu_0^{LF}(\underline{\rho}) = S^{-1}\int d^2\underline{s}\ \exp(ik\underline{\rho}\underline{s})\mathrm{circ}(\underline{s})(1-\underline{s}^2)^{-1}J(\underline{s}) \quad . \tag{5.77}$$

In principle this relation enables us to recover the low frequency part $\mu_0^{LF}(\underline{\rho})$ of the degree of coherence $\mu_0(\underline{\rho})$ from the radiant intensity $J(\underline{s})$. We have, however, to bear in mind that the inverse relation (5.77) is useful only if the source is already known to be quasistationary. The same source type allows the determination of the low-frequency part of the intensity profile $I_0(\underline{\rho}')$ by Fourier inversion of the "symmetrically scanned" [5.143] far-zone coherence function $W_r(\underline{s}_1,\underline{s}_2=-\underline{s}_1)$ with the angular correlation $(\underline{s}'=2\underline{s}_1)$

$$F(\underline{s}',0) = \int d^2\underline{\rho}\,\mu_0(\underline{\rho})\int d^2\underline{\rho}'\ \exp(-ik\underline{s}'\underline{\rho}')I_0(\underline{\rho}') \quad . \tag{5.78}$$

### 5.4.3 Bessel-Correlated Sources

We illustrate the relationship between the degree of coherence $\mu_0(\underline{\rho})$ in the source plane, the angular spectrum $F(\underline{s})$, and the radiant intensity $J(\underline{s})$ in terms of the *quasistationary* isotropic Bessel-correlated source (5.54,57) leading to [5.29,116]

$$F(\underline{s})/F(0) = \begin{cases}(1-s^2)^{n/2-1} & , \quad 0 \leqq s \leqq 1 \\ 0 & , \quad s \geqq 1\end{cases} \tag{5.79}$$

and

$$J(\underline{s})/J(0) = \cos^n\vartheta \quad . \tag{5.80}$$

We display the functions (5.57,79,80) in Fig.5.3 for a variety of values of the parameter n between 0.1 and 40. This parameter controls the degree of coherence, the angular spectrum, and the directionality of the source. We observe that a high degree of spatial coherence (large n) goes along with increasing importance of the low angular frequencies ($s \ll 1$) and pronounced directionality. We recall that the case $n = 1$ corresponds to the blackbody or Lambertian source (dashed curves in Fig.5.3).

Numerical evaluation [5.119] of the *Schell's* Bessel-correlated source (5.50,57) with circular aperture of radius a shows that the quasistationary approximation is fair provided that $ak \gtrsim 100$ and $ak \gtrsim n \gtrsim 2$. The latter condition means that 1) the width of the central peak of $\mu_0(\underline{\rho})$ should be small compared to the source diameter and 2) the oscillatory tail of $\mu_0(\underline{\rho})$ should decay sufficiently rapidly as $k\rho \to \infty$. The blackbody with small aperture, $ak \lesssim 100$, is expected to show sensible deviations from the Lambertian radiant intensity due to the slow decay of $(k\rho)^{-1}\sin k\rho$.

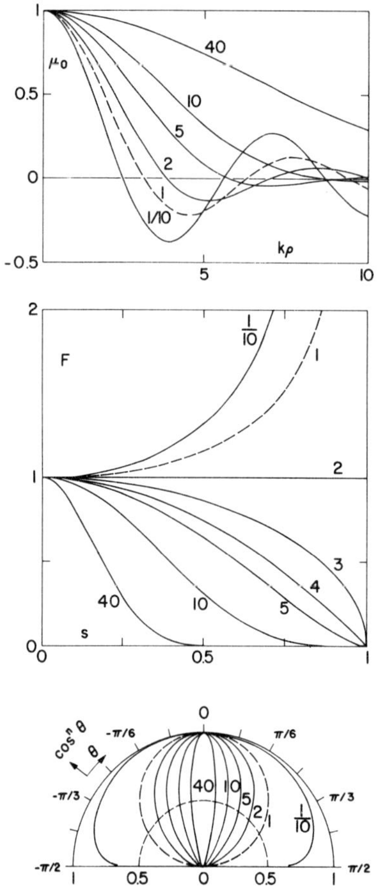


Fig. 5.3. Properties of Bessel-correlated sources: degree of coherence (above), angular spectrum (middle), and radiant intensity (below) (adapted from [5.29])

### 5.4.4 Gauss-Correlated Sources

The radiometry of partially coherent planar sources of arbitrary size is best illustrated in terms of the isotropic Gauss-correlated *Schell's* source (5.50,53,56), viz.,

$$W_0(\underline{\rho}_1,\underline{\rho}_2) \propto \exp\left[-\left(|\underline{\rho}_1|^2+|\underline{\rho}_2|^2\right)/4a^2-|\underline{\rho}_1-\underline{\rho}_2|^2/2b^2\right] \quad . \tag{5.81}$$

In their theory of the fluctuations of radiation scattered by a deep random phase screen, JAKEMAN and PUSEY [5.133] and JAKEMAN and McWHIRTER [5.134] have derived the far-zone properties of the source (5.81). BALTES and STEINLE [5.120-123] have established the corresponding radiometric relations. The Gauss-correlated quasistationary source studied by CARTER and WOLF [5.114,115] is a special case of the above model, namely the limit $a \gg b$. We recall that the effective beam width a controls the size of the source, whereas the effective coherence length b controls its degree of coherence.

By virtue of (5.68,69,72,73) the source (5.81) has the reduced radiant intensity

$$J(\underline{s})/J(0) = \cos^2\vartheta \exp\left[-\sin^2\vartheta/2(\Delta I)^2\right] \tag{5.82}$$

and the degree of angular coherence

$$\mu(\underline{s}_1,\underline{s}_2) = \exp\left[-|\underline{s}_1-\underline{s}_2|^2/2(\Delta\mu)^2\right] \tag{5.83}$$

with the "beam spread" [5.121] or "apparent beam width" [5.134]

$$\Delta I = k^{-1}(b^{-2}+a^{-2}/4)^{1/2} \tag{5.84}$$

and the "coherence angle"

$$\Delta\mu = a^{-1}b\ \Delta I = (ak)^{-1}\left[1+(b/2a)^2\right]^{1/2} \quad . \tag{5.85}$$

We observe that the parameters b and 2a occur in the radiant intensity (but not the degree of angular coherence) in a symmetric way. From pronounced directionality we thus conclude that both the source diameter a *and* the coherence length b are large compared with $k^{-1}$. The parameters a and b can, in principle, be determined from measuring the radiant intensity (5.82) and/or the degree of angular coherence (5.83); see also Section 6.3.1 and [5.121]. In the quasistationary limit $a \gg b$ we have $\Delta I \to (bk)^{-1}$ and $\Delta\mu \to (ak)^{-1}$ in agreement with the general reciprocity relation presented in Section 5.3.3. Both variances (5.84) and (5.85) show the scaling factor $[1+(b/2a)^2]^{1/2}$ with respect to the limit $a \gg b$. The same scaling behavior occurs in the scattering experiments and numerical calculations due to BERTOLOTTI et al. [5.139].

### 5.4.5 An Application: Coherence of Thermionic Sources

FERWERDA and VAN HEEL [5.124-126] have shown that the virtual electron source (virtual cathode) is not completely incoherent, as was traditionally assumed, but exhibits a nonzero coherence length. The initial Maxwell-Boltzmann distribution and the accelerating voltage U lead to the effective radiant intensity

$$J(\underline{s}) = J(0)\exp[-(eU/KT)\sin^2\vartheta] \tag{5.86}$$

of the virtual cathode, where e denotes the elementary charge. Typically one has $U = 3\cdot 10^5$V and $T = 3000$K, i.e., very high directionality.

Using the quasistationary model (5.54) and the inverse relation (5.77), we obtain a Gauss-correlated virtual source, viz.,

$$\mu_0(\underline{\rho}) = \exp\left[-(KT/4eU)k^2\rho^2\right] \quad , \tag{5.87}$$

where k now denotes the de Broglie wavenumber of the accelerated electrons. The coherence length b as defined by (5.56) reads

$$b = k^{-1}(2eU/KT)^{1/2} = \lambda_{th} \tag{5.88}$$

with $\lambda_{th}$ denoting the thermal wavelength. For a cathode temperature of $T = 3000$ K we find $b \simeq 3$ Å, a number that is substantially larger than the wavelength of $10^5$ eV electrons, 0.037 Å. On the other hand, we learn from the result (5.88) that the underlying physical source (thermal electrons) shows poor spatial coherence, the coherence length being of the order of the wavelength.

## 5.5 Radiation Efficiency

Let us finally illustrate the radiometric quantities of Section 5.3 using the model sources of Section 5.4. The total hemispherical flux radiated by a source depends not only on the beam profile $I_0(\underline{\rho}_1)$, but also on the degree of coherence $\mu_0(\underline{\rho})$. The maximum possible flux is obtained when both beamwidth and coherence length are very large compared with the wavelength [5.121-123]. Reduction of either beamwidth or coherence length leads to a decrease of the flux. This situation is described in terms of the "radiation efficiency" [5.115] or "transfer factor" [5.121].

### 5.5.1 Radiance of Model Sources

Rewriting the radiance defined by (5.34) as

$$B^{(1)}(\underline{\rho}',\underline{s}) \propto \cos\vartheta \int d^2\underline{\rho}\ \exp(-ik\underline{s}\underline{\rho})W_0(\underline{\rho}'+\underline{\rho}/2,\underline{\rho}'-\underline{\rho}/2) \tag{5.89}$$

we find [5.121]

$$B^{(1)}(\underline{\rho}',\underline{s}) \propto \exp(-|\underline{\rho}'|^2/2a^2)(\Delta I)^{-2}\cos\vartheta\ \exp[-\sin^2\vartheta/2(\Delta I)^2] \tag{5.90}$$

for the *Gauss-correlated* source (5.81), where $\Delta I$ is given by (5.84). The expression (5.90) is also the leading term in the radiance of a deep random phase screen illuminated by a coherent Gaussian beam [5.123]. In the quasistationary limit [5.114], $(\Delta I)^{-2}$ is replaced by $(bk)^2$. The radiance of the *Bessel-correlated* quasistationary source (5.57) is proportional to $(\cos\vartheta)^{n-1}$.

### 5.5.2 Emittance and Radiation Efficiency

The emittance

$$E^{(1)}(\underline{\rho}') = \int d\Omega \cos\vartheta \, B^{(1)}(\underline{\rho}',\underline{s}) \tag{5.91}$$

corresponding to (5.89) can be expressed as

$$E^{(1)}(\underline{\rho}') = C^{(1)} \, I_0(\underline{\rho}') \tag{5.92}$$

with the *radiation efficiency*

$$C^{(1)} \equiv \int d\Omega \cos\vartheta \, \tilde{B}^{(1)}(\underline{s}) \quad , \tag{5.93}$$

provided that the radiance (5.89) factorizes, viz.,

$$B^{(1)}(\underline{\rho}',\underline{s}) \propto I_0(\underline{\rho}') \, \tilde{B}^{(1)}(\underline{s}) \quad . \tag{5.94}$$

This condition is fulfilled by any quasistationary source (5.54) and, moreover, by the special Schell's source (5.81) as is seen from (5.90).

### 5.5.3 Examples

The radiation efficiency of the *Gauss-correlated* source with Gaussian beam profile see (5.81) reads [5.121-123]

$$C^{(1)} = \frac{1}{3} (\Delta I)^{-2} \, M\left[1,5/2,-(\Delta I)^{-2}/2\right] \tag{5.95}$$

with M denoting the confluent hypergeometric function. We plot this result in Figure 5.4. We recall that $(\Delta I)^{-2} = k^2/(b^{-2}+a^{-2}/4)$. Thus $C^{(1)} \to 1$ provided that both $ak \gg 1$ and $bk \gg 1$, whereas $C^{(1)} \ll 1$ if either $ak \lesssim 1$ or $bk \lesssim 1$. The relation (5.95) comprises the earlier results of [5.114] in the quasistationary limit $a \gg b$. The

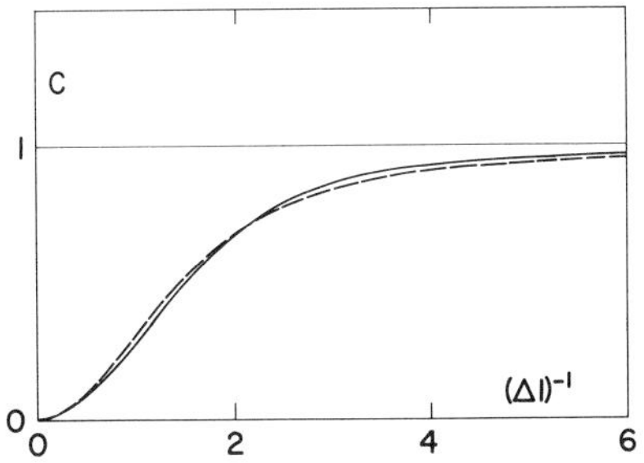


Fig. 5.4. Radiation efficiency (full curve) and fluctuation efficiency (dashed curve) of Gauss-correlated sources

opposite limit, $b \gg a$, corresponds to the classical diffraction result for the propagation of a coherent Gaussian beam. In deriving (5.95), the integration over the solid angle elements $d\Omega = \sin\vartheta \, d\vartheta \, d\varphi$ is restricted to real angles $\vartheta$, i.e., nonevanescent waves. Extending the integration to complex angles leads to an additional imaginary part of the radiation efficiency [5.121], which becomes rapidly negligible with increasing beam width a and coherence length b. The corresponding radiation efficiency of the deep random screen is discussed in [5.123].

We finally mention the radiation efficiency of the *Bessel-correlated* quasistationary source [5.121], viz.,

$$C^{(1)} = n(n+1)^{-1} \quad , \tag{5.96}$$

with $C^{(1)} \to 1$ in the coherent limit $n \to \infty$. In particular, the Lambertian source (n=1) generates one half of the flux produced by a spatially coherent source [5.115,121].

## 5.6 Second-Order Radiometry

Measurement of second-order correlation (intensity interferometry) is a powerful tool for obtaining significant features of direct and indirect optical sources [5.23-30]. Being based on the first-order coherence function (5.11), the radiometry of partially coherent sources developed in Sections 5.3-5 cannot account for second-order properties like the fluctuation of the radiant intensity and the related radiant intensity autocorrelation. Only recently BALTES and coworkers [5.123,152-154] proposed further radiometric concepts with the aim of incorporating second-order coherence into the field of radiometry. The new concepts are summarized below.

### 5.6.1 Radiant Intensity Fluctuation and Autocorrelation

Second-order radiometry is based on the four-point correlation function

$$W^{(2)}(\underline{r}_1,\underline{r}_2,\underline{r}_3,\underline{r}_4) \equiv \langle u^*(\underline{r}_1)u^*(\underline{r}_2)u(\underline{r}_3)u(\underline{r}_4)\rangle \quad . \tag{5.97}$$

Let us denote by $W_0^{(2)}(\underline{\varrho}_1,\underline{\varrho}_2,\underline{\varrho}_3,\underline{\varrho}_4)$ and $W_r^{(2)}(\underline{s}_1,\underline{s}_2,\underline{s}_3,\underline{s}_4)$ the corresponding source-plane and far-zone correlations analogous to (5.49) and (5.64). The *radiant intensity autocorrelation* reads

$$\begin{aligned}\langle J(\underline{s}_1)J(\underline{s}_2)\rangle \equiv r^4 W_r^{(2)}(s_1,s_2,s_2,s_1) &= (k/2\pi)^4 \cos^2\vartheta_1 \cos^2\vartheta_2 \int d^2\underline{\varrho} \int d^2\underline{\varrho}'' \\ &\exp\left[-ik(\underline{s}\underline{\varrho}+\underline{s}'\underline{\varrho}'')\right] \int d^2\underline{\varrho}''' \int d^2\underline{\varrho}' \, W_0^{(2)}(\varrho_1,\varrho_2,\varrho_3,\varrho_4)\end{aligned} \tag{5.98}$$

with $\underline{s}$, $\underline{s}'$ as defined by (5.65), but with the transformation [5.133,134]

$$\underline{\rho} = \underline{\rho}_1 + \underline{\rho}_2 - \underline{\rho}_3 - \underline{\rho}_4 \quad , \quad \underline{\rho}' = (\underline{\rho}_1+\underline{\rho}_2+\underline{\rho}_3+\underline{\rho}_4)/4$$
$$\underline{\rho}'' = (\underline{\rho}_1-\underline{\rho}_2+\underline{\rho}_3-\underline{\rho}_4)/2 \qquad \underline{\rho}''' = (\underline{\rho}_1-\underline{\rho}_2-\underline{\rho}_3+\underline{\rho}_4)/2 \quad , \tag{5.99}$$

instead of (5.51,55). The *radiant intensity fluctuation* is obtained from (5.98) in the case $\underline{s}_1 = \underline{s} = \underline{s}_2$, viz.,

$$\left\langle J^2(\underline{s})\right\rangle \equiv r^4 W_r^{(2)}(\underline{s},\underline{s},\underline{s},\underline{s}) = (k/2\pi)^4 \cos^4\vartheta \int d^2\underline{\rho}\, \exp(-ik\underline{s}\underline{\rho}) \int d^2\underline{\rho}'' \int d^2\underline{\rho}''' \int d^2\underline{\rho}'\, W_0^{(2)}(\dots) \quad . \tag{5.100}$$

## 5.6.2 Second-Order Radiometric Quantities

One feasible way of defining the *second-order radiance* $B^{(2)}(\underline{\rho}',\underline{s})$ is [5.123,152-154]

$$B^{(2)}(\underline{\rho}',\underline{s}) \equiv (k/2\pi)^4 \cos^2\vartheta \int d^2\underline{\rho}\, \exp(-ik\underline{s}\underline{\rho}) \int d^2\underline{\rho}'' \int d^2\underline{\rho}'''\, W_0(\dots) \quad . \tag{5.101}$$

This definition allows the relation

$$\left\langle J^2(\underline{s})\right\rangle = \cos^2\vartheta \int d^2\underline{\rho}'\, B^{(2)}(\underline{\rho}',\underline{s}) \tag{5.102}$$

analogous to (5.10b). A second-order quantity analogous to the flux (5.10) is, e.g., the *total hemispherical fluctuation*

$$\left\langle \Phi^2\right\rangle \equiv \int d\Omega \left\langle J^2(\underline{s})\right\rangle = \int d\Omega \int d^2\underline{\rho}' \cos^2\vartheta\, B^{(2)}(\underline{\rho}',\underline{s}) \quad . \tag{5.103}$$

This definition corresponds to angular scanning with subsequent mathematical integration of the measured radiant intensity fluctuation over the hemisphere. The quest for an expression of (5.103) in the form

$$\left\langle \Phi^2\right\rangle = \int d^2\underline{\rho}'\, E^{(2)}(\underline{\rho}') \tag{5.104}$$

leads to the definition of the *second-order emittance*, viz.,

$$E^{(2)}(\underline{\rho}') \equiv \int d\Omega \cos^2\vartheta\, B^{(2)}(\underline{\rho}',\underline{s}) = \left[3k^4/2(2\pi)^{5/2}\right] \int d^2\underline{\rho}\, (k\rho)^{-5/2} J_{5/2}(k\rho) \int d^2\underline{\rho}'' \int d^2\underline{\rho}'''\, W_0(\dots) \quad . \tag{5.105}$$

This quantity can be written as

$$E^{(2)}(\underline{\rho}') = C^{(2)} \left\langle I_0^2(\underline{\rho}')\right\rangle \tag{5.106}$$

where $C^{(2)}$ denotes the *fluctuation efficiency* and $\langle I_0^2(\underline{\rho}')\rangle \equiv W_0^{(2)}(\underline{\rho}',\underline{\rho}',\underline{\rho}',\underline{\rho}')$, provided that $B^{(2)}(\underline{\rho}',\underline{s})$ factorizes, viz.,

$$B^{(2)}(\underline{\rho}',\underline{s}) = \tilde{B}^{(2)}(\underline{s})\langle I_0^2(\underline{\rho}')\rangle \quad . \tag{5.107}$$

The condition (5.107) is fulfilled by quasistationary sources where $W_0$ depends only on differences of the $\underline{\rho}_j$.

### 5.6.3 An Example: Gauss-Correlated Chaotic Source

We consider the chaotic source with Gaussian correlation (coherence length b) and Gaussian intensity profile (beam width a), i.e.,

$$W_0^{(2)}(\underline{\rho}_1,\underline{\rho}_2,\underline{\rho}_3,\underline{\rho}_4) \propto \exp\left(-\sum_{j=1}^{4}|\underline{\rho}_j|^2/4a^2\right)\left[\exp\left(-|\underline{\rho}_1-\underline{\rho}_4|^2/2b^2\right)\exp\left(-|\underline{\rho}_2-\underline{\rho}_3|^2/2b^2\right)\right.$$
$$\left.+\exp\left(-|\underline{\rho}_1-\underline{\rho}_3|^2/2b^2\right)\exp\left(-|\underline{\rho}_2-\underline{\rho}_4|^2/2b^2\right)\right] \quad . \tag{5.108}$$

The corresponding second-order radiance reads

$$B^{(2)}(\underline{\rho}',\underline{s}) \propto a^2(\Delta I)^{-4}\exp(-\underline{\rho}'/a^2)\cos^2\vartheta\,\exp\left[-\sin^2\vartheta/(\Delta I)^2\right] \tag{5.109}$$

with $\Delta I$ as given by (5.84). We observe that $B^{(2)}$ is related to the square of $B^{(1)}$ in this case [see (5.90)]. The pertinent fluctuation efficiency reads

$$C^{(2)} = \frac{4}{5}a^2(\Delta I)^{-4}\,M\left[1,7/2,-(\Delta I)^{-2}\right] \tag{5.110}$$

with M denoting the confluent hypergeometric function. In Figure 5.4 we plot the normalized fluctuation efficiency $[(\Delta I)^2/2a^2]C^{(2)}$. The analogous results for a model of the deep random phase screen can be found in [5.154].

More research has to be done to develop the involved second-order radiometry of nonchaotic sources (see Chap.6) and, in particular, the second-order relationship analogous to the Van Cittert-Zernike theorem.

## References


5.1 J. Geist: Opt. Eng. *15*, 537-540 (1976) (review)
5.2 G. Bauer: *Measurement of Optical Radiation* (Focal Press, New York 1965)
5.3 J.W.T. Walsh: *Photometry* (Dover, New York 1965)
5.4 E.M. Sparrow, R.D. Cess: *Radiation Heat Transfer* (Brooks-Cole, Belmont 1966)
5.5 J.G. Calvert, J.N. Pitts: *Photochemistry* (Wiley and Sons, New York 1966) pp.780-788

5.6 S. Chandrasekhar: *Radiative Transfer* (Dover, New York 1960)
5.7 G. Kortüm: *Reflexionsspektroskopie* (Springer, Berlin, Heidelberg, New York 1969)
5.8 H.G. Hecht: J. Res. Nat. Bur. Stand. *80A*, 567-583 (1976) (review)
5.9 A. Hadni: *Essentials of Modern Physics Applied to the Study of the Infrared* (Pergamon Press, New York 1967) p.2
5.10 Y. Le Grand: *Light, Color and Vision* (Wiley and Sons, New York 1957) pp.66-72
5.11 V.K. Zworykin, E.G. Ramberg: *Photoelectricity* (Wiley and Sons, New York 1949)
5.12 D.B. Judd: *Color in Business, Science and Industry* (Wiley and Sons, New York 1952) pp.83-94
5.13 G. Wyszecki, W.S. Stiles: *Color Science* (Wiley and Sons, New York 1967) pp.228-370
5.14 R.A. Smith, F.E. Jones, R.P. Chasmar: *The Detection and Measurements of Infrared Radiation* (Clarendon Press, Oxford 1968)
5.15 K. Codling, R.P. Madden: J. Appl. Phys. *36*, 380-387 (1965)
5.16 W.R. Ott, P. Fieffe-Prevost, W.L. Wiese: Appl. Opt. *12*, 1618-1629 (1973)
5.17 C. Fröhlich, J. Geist, J.M. Kendall, R.M. Marchgraber: Sol. Eng. *14*, 157-166 (1973)
5.18 J. Geist, L.B. Schmidt, W.E. Case: Appl. Opt. *12*, 2773-2776 (1973)
5.19 J. Geist, B. Steiner, R. Schaefer, E. Zalewski, A. Corrons: Appl. Phys. Lett. *26*, 309-311 (1975)
5.20 W.R. Blevin, B. Steiner: Metrologia *11*, 97-104 (1975)
5.21 R.T. Birge: Rev. Mod. Phys. *13*, 233-239 (1941)
5.22 W.R. Blevin, W.J. Brown: Metrologia *7*, 15-29 (1971)
5.23 J.R. Klauder, E.C.G. Sudarshan: *Fundamentals of Quantum Optics* (Benjamin, New York, Amsterdam 1968)
5.24 F.T. Arecchi, V. Degiorgio: "Measurements of the Statistical Properties of Optical Fields, in *Laser Handbook*, ed. by F.T. Arecchi, E.O. Schulz-Dubois, Vol.1 (North-Holland, Amsterdam, New York, Oxford 1972) Chap.A5, pp.191-264
5.25 H.Z. Cummins, E.R. Pike (eds.): *Photon Correlation and Light Beating Spectroscopy*, Nato Advanced Study Institute Lectures, Capri 1973 (Plenum Press, New York, London 1974)
5.26 R. Hanbury Brown: *The Intensity Interferometer* (Taylor and Francis, London 1974)
5.27 B. Crossignani, P. DiPorto, M. Bertolotti: *Statistical Properties of Scattered Light* (Academic Press, New York 1975)
5.28 J.C. Dainty (ed.): *Laser Speckle and Related Phenomena*, Topics in Applied Physics, Vol.9 (Springer, Berlin, Heidelberg, New York 1975)
5.29 H.P. Baltes: Appl. Phys. *12*, 221-244 (1977) (review)
5.30 A. Zardecki, D. Delisle: Opt. Acta *24*, 241-259 (1977) (review)
5.31 R.C. Bourret: Nuovo Cimento *18*, 347-356 (1960)
5.32 H.P. Baltes, E.R. Hilf: *Spectra of Finite Systems* (Bibliographisches Institut, Zürich 1976)
5.33 H.P. Baltes: Infrared Phys. *16*, 1-8 (1976) (review)
5.34 A. Walther: J. Opt. Soc. Am. *58*, 1256-1259 (1968)
5.35 E. Wolf: Phys. Rev. D *13*, 869-886 (1976)
5.36 H.P. Baltes: "On the Validity of Kirchhoff's Law of Heat Radiation for a Body in a Nonequilibrium Environment", in *Progress in Optics*, ed. by E. Wolf, Vol.13 (North-Holland, Amsterdam, Oxford and American Elsevier, New York 1976) pp.1-25
5.37 A. Kano, E. Wolf: Proc. Phys. Soc. *80*, 1273-1276 (1962)
5.38 J. Sarfatt: Nuovo Cimento *27*, 1119-1129 (1963)
5.39 C.L. Mehta, E. Wolf: Phys. Rev. *134*, A1143-1149 (1964)
5.40 C.L. Mehta, E. Wolf: Phys. Rev. *134*, A1149-1153 (1964)
5.41 E. Fox Keller: Phys. Rev. *139*, B202-211 (1965)
5.42 L. Mandel, E. Wolf: Rev. Mod. Phys. *37*, 231-287 (1965) Sect.5.7 (review)
5.43 C.L. Mehta, E. Wolf: Phys. Rev. *161*, 1328-1334 (1967)
5.44 R. Fürth: Proc. Roy. Soc. Edinburgh *67 A*, 289-302 (1967)
5.45 J.H. Eberly, A. Kujawski: Phys. Rev. *155*, 10-19 (1967)
5.46 A. Kujawski: Acta Phys. Pol. *34*, 957-966 (1968)
5.47 I. Brevik, E. Suhonen: Phys. Norv. *3*, 135-150 (1968)
5.48 I. Brevik, E. Suhonen: Nuovo Cimento *60*, *B* 141-157 (1969)
5.49 I. Brevik, E. Suhonen: Nuovo Cimento *65*, *B* 187-207 (1970)
5.50 R.K. Nesbet: Phys. Rev. A *4*, 259-264 (1971)

5.51 R.K. Nesbet: Phys. Rev. Lett. *27*, 553-556 (1971)
5.52 F.R. Nash, J.P. Gordon: Phys. Rev. A *12*, 2472-2486 (1975)
5 53 J.P. Gordon: Phys. Rev. A *12*, 2487-2497 (1975)
5.54 H.W. Lee, P. Stehle: Phys. Rev. Lett. *36*, 277-279 (1976)
5.55 T.H. Boyer: Phys. Rev. *174*, 1764-1776 (1968)
5.56 W. Lukosz: Z. Phys. *262*, 327-348 (1973)
5.57 R. Balian, B. Duplantier: Electromagnetic waves near perfect conductors. II. Casimir effect. Ann. Phys. (N.Y.) *112* (1978)
5.58 D. Polder, M. van Hove: Phys. Rev. B *4*, 3303-3314 (1971)
5.59 C.M. Hargreaves: "Radiative Transfer Between Closely Spaced Bodies", Proefschrift, Univ. Leiden (1973)
5.60 R. Balian, C. Bloch: Ann. Phys. (N.Y.) *64*, 271-307 (1971); *84*, 559-563 (1974) errata
5.61 H.P. Baltes, F.K. Kneubühl: Helv. Phys. Acta *45*, 481-529 (1972)
5.62 H.P. Baltes: Phys. Rev. A *6*, 2252-2257 (1972)
5.63 H.P. Baltes: J. Math. Phys. *18*, 1275-1276 (1976)
5.64 R. Balian, B. Duplantier: Electromagnetic waves near perfect conductors I. Multiple scattering expansions. Distribution of modes. Ann. Phys. (N.Y.) *104* (1977)
5.65 K. Case, S.C. Chiu: Phys. Rev. A *1*, 1170-1174 (1970)
5.66 H.P. Baltes: Appl. Phys. *1*, 39-43 (1973)
5.67 W. Eckhardt: Opt. Commun. *14*, 95-98 (1975)
5.68 H.P. Baltes: J. Phys. (Paris) *7*, C2, 151-156 (1977) (review)
5.69 H.P. Baltes, E.R. Hilf, M. Pabst: Appl. Phys. *3*, 21-29 (1974); *5*, 83 (1974) errata
5.70 B. Steinle, H.P. Baltes, M. Pabst: Phys. Rev. A *12*, 1519-1524 (1975)
5.71 H.P. Baltes, B. Steinle, M. Pabst: Phys. Rev. A *13*, 1866-1873 (1976)
5.72 G.S. Agarwal: Phys. Rev. A *11*, 230-242 (1975)
5.73 G.S. Agarwal: Phys. Rev. A *11*, 253-264 (1975)
5.74 G.S. Agarwal: Phys. Rev. A *12*, 1974-1986 (1975)
5.75 W. Eckhardt: Z. Phys. B *23*, 213-219 (1976)
5.76 W. Eckhardt: Z. Phys. B *26*, 291-297 (1977)
5.77 E. Wolf: Proc. Phys. Soc. *80*, 1269-1272 (1962)
5.78 H.M. Nussenzveig: J. Math. Phys. *8*, 561-572 (1967)
5.79 R.J. Glauber: Phys. Rev. *131*, 2766-2788 (1963)
5.80 C.L. Mehta, E. Wolf: Phys. Rev. *157*, 1188-1197 (1967)
5.81 L. Mandel, E. Wolf: J. Opt. Soc. Am. *66*, 529-535 (1976)
5.82 M. Planck: *The Theory of Heat Radiation* (Dover, New York 1959)
5.83 M.J. Beran, G.B. Parrent, Jr.: *Theory of Partial Coherence* (Prentice Hall, Englewood Cliffs 1964)
5.84 C.L. Mehta, E. Wolf, A.P. Balachandran: J. Math. Phys. *7*, 133-138 (1966)
5.85 H.S. Green, E. Wolf: Proc. Phys. Soc. A *66*, 1129-1137 (1953)
5.86 J. Focke: Opt. Acta *4*, 124-126 (1957)
5.87 A. Walther: J. Opt. Soc. Am. *57*, 639-644 (1967)
5.88 A. Walther: Am. J. Phys. *34*, 521-525 (1966)
5.89 M. Kline, I.W. Kay: *Electromagnetic Theory and Geometrical Optics*, Sect.XII-8 (Interscience, New York 1965)
5.90 E.W. Marchand, E. Wolf: J. Opt. Am. *64*, 1219-1226 (1974)
5.91 M. Abramowitz, I.A. Stegun (eds.): *Handbook of Mathematical Functions*, N.B.S. Appl. Math. Series 55 (U.S. Department of Commerce, Washington DC 1964) p.485, Eq.11.4.10
5.92 E.W. Marchand, E. Wolf: J. Opt. Soc. Am. *64*, 1273-1274 (1974)
5.93 A. Walther: J. Opt. Soc. Am. *64*, 1275 (1974)
5.94 A. Walther: J. Opt. Soc. Am. *63*, 1622-1623 (1973)
5.95 E. Wigner: Phys. Rev. *40*, 749-759 (1932)
5.96 P.H. Van Cittert: Physica *1*, 201-210 (1934)
5.97 F. Zernike: Physica *5*, 785-795 (1938)
5.98 P.H. Van Cittert: Physica *6*, 1129-1138 (1939)
5.99 P.H. Van Cittert: "Physical Optics", in *Textbook of Physics*, ed. by R. Kronig (Pergamon Press, New York 1959) Chap.5, pp. 376-475

5.100 D. Middleton: *Statistical Communication Theory* (McGraw-Hill, New York 1960) Sect.8.1
5.101 M. Born, E. Wolf: *Principles of Optics*, 4th ed. (Pergamon Press, Oxford 1970)
5.102 E.L. O'Neill: *Introduction to Statistical Optics* (Addison-Wesley, Reading, Mass. 1963)
5.103 A.C. Schell: "The Multiple Plate Antenna", Doctoral Dissertation, Massachusetts Institute of Technology (1961) Sect.7.5
5.104 R.A. Shore: "Partially Coherent Diffraction by a Circular Aperture", in *Electromagnetic Theory and Antennas*, ed. by E.C. Jordan, Pt.2 (Pergamon Press, London 1963) pp.787-795
5.105 R.A. Shore, B.J. Thompson, R.E. Whitney: J. Opt. Soc. Am. *56*, 733-738 (1966)
5.106 A.K. Jaiswal, C.L. Mehta: Opt. Commun. *5*, 50-52 (1972)
5.107 A.K. Jaiswal, G.P. Agrawal, C.L. Mehta: Nuovo Cimento B *15*, 295-307 (1973)
5.108 A. Walther: Am. J. Phys. *36*, 808-816 (1967)
5.109 E.W. Marchand, E. Wolf: J. Opt. Soc. Am. *62*, 379-385 (1972)
5.110 E.W. Marchand, E. Wolf: Opt. Commun. *6*, 305-308 (1972)
5.111 E. Wolf, W.H. Carter: Opt. Commun. *13*, 205-209 (1975)
5.112 W.H. Carter, E. Wolf: J. Opt. Soc. Am. *65*, 1067-1071 (1975)
5.113 E. Wolf, W.H. Carter: Opt. Commun. *16*, 297-302 (1976)
5.114 W.H. Carter, E. Wolf: J. Opt. Soc. Am. *67*, 785-796 (1977)
5.115 E. Wolf, W.H. Carter: "On the Radiation Efficiency of Quasi-Homogeneous Sources of Different Degrees of Spatial Coherence", in *Proc. 4th Rochester Conf. on Coherence and Quantum Optics*, June 8-10, 1977, ed. by L. Mandel, E. Wolf (Plenum Press, New York 1978)
5.116 H.P. Baltes, B. Steinle, G. Antes: Opt. Commun. *18*, 242-246 (1976)
5.117 G. Antes, H.P. Baltes, B. Steinle: Helv. Phys. Acta *49*, 759-761 (1976)
5.118 B. Steinle, H.P. Baltes: Helv. Phys. Acta *49*, 793-795 (1976)
5.119 B. Steinle, H.P. Baltes: J. Opt. Soc. Am. *67*, 241-247 (1977)
5.120 H.P. Baltes, B. Steinle: Lett. Nuovo Cimento *18*, 313-318 (1977)
5.121 H.P. Baltes, B. Steinle: Nuovo Cimento B *41*, 428-440 (1977)
5.122 B. Steinle, H.P. Baltes: Helv. Phys. Acta *50*, 664-666 (1977)
5.123 H.P. Baltes, B. Steinle, G. Antes: "Radiometric and Correlation Properties of Bounded Planar Sources", in *Proc. 4th Rochester Conf. on Coherence and Quantum Optics*, June 8-10, 1977, ed. by L. Mandel, E. Wolf (Plenum Press, New York 1978) pp.431-441
5.124 H.A. Ferwerda: Optik *45*, 411-426 (1976)
5.125 H.A. Ferwerda, M.G. van Heel: Optik *47*, 357-362 (1977)
5.126 H.A. Ferwerda, M.G. van Heel: "Determination of Coherence Length from Directionality", in *Proc. 4th Roch. Conf. on Coherence and Quantum Optics*, June 8-10, 1977, ed. by L. Mandel, E. Wolf (Plenum Press, New York 1978)
5.127 P.W. Hawkes: Optik *47*, 453-467 (1977)
5.128 G. Ross: Phil. Trans. Roy. Soc. (London) *268*, 177-200 (1970)
5.129 T. Asakura, I. Akamatsu: Opt. Acta *19*, 749-763 (1972)
5.130 T. Asakura, I. Akamatsu: Opt. Acta *20*, 129-136 (1973)
5.131 H. Fujii, T. Asakura: Appl. Phys. *3*, 121-129 (1974)
5.132 R.S. Sirohi, V.R. Mohan: Opt. Acta *22*, 207-210 (1975)
5.133 E. Jakeman, P.N. Pusey: J. Phys. A *8*, 369-391 (1975)
5.134 E. Jakeman, J.G. McWhirter: J. Phys. A *9*, 785-797 (1976)
5.135 T. Asakura, H. Fujii, K. Murata: Opt. Acta *19*, 273-290 (1972)
5.136 H. Fujii, T. Asakura: Optik *39*, 99-117 (1973)
5.137 H. Fujii, T. Asakura: Optik *39*, 284-302 (1974)
5.138 P.N. Pusey, E. Jakeman: J. Phys. A *8*, 392-410 (1975)
5.139 M. Bertolotti, F. Scudieri, S. Verginelli: Appl. Opt. *15*, 1842-1844 (1976)
5.140 M. Bertolotti, F. Scudieri, A. Ferrari, D. Apostol: Appl. Opt. *15*, 2468-2470 (1976)
5.141 M. Bertolotti, F. Scudieri: "Spatial Coherence of Light Scattered by Liquid Crystals", in *Proc. 4th Rochester Conf. on Coherence and Quantum Optics*, June 8-10, 1977, ed. by L. Mandel, E. Wolf (Plenum Press, New York 1978)
5.142 W.H. Carter: Appl. Phys. *16*, 558-563 (1977)
5.143 S. Wadaka, T. Sato: J. Opt. Soc. Am. *66*, 145-147 (1976)
5.144 H.A. Ferwerda: Opt. Commun. *19*, 54-56 (1976)

5.145 K. Dutta, J.W. Goodman: J. Opt. Soc. Am. *67*, 796-803 (1977)
5.146 D.J. Carpenter, C. Pask: J. Opt. Soc. Am. *67*, 115-117 (1977)
5.147 D.J. Carpenter, C. Pask: Opt. Acta *23*, 279-286 (1976)
5.148 C. Pask: Opt. Acta *24*, 235-240 (1977)
5.149 J.C. Leader: The generalized partial coherence of a radiation source and its far-field. Opt. Acta *25*, 395-413 (1978)
5.150 J.C. Leader: Equivalent source coherence of laser-illuminated rough surfaces J. Opt. Soc. Am. *66*, 183 (1976) and submitted to Opt. Acta
5.151 J.R. Shewell, E. Wolf: J. Opt. Soc. Am. *58*, 1596-1603 (1967)
5.152 H.P. Baltes: Lett. Nuovo Cimento *20*, 87-90 (1977)
5.153 H.P. Baltes, B. Steinle: J. Opt. Soc. Am. *67*, 1366 (1977)
5.154 H.P. Baltes, B. Steinle: Fluctuating sources and second-order radiometry. Nuovo Cimento B *44*, 423-441 (1978)

# 6. Statistical Features of Phase Screens from Scattering Data

A. Zardecki

With 6 Figures

A random phase screen (RPS) is a system which retards the phase of an incident electromagnetic field by a randomly varying, position-dependent amount. This concept has been useful in describing a variety of effects accompanying the propagation of electromagnetic waves. We mention but a few typical examples. When passing through the terrestrial ionosphere [6.1-3], the radio waves from radio stars may suffer irregular phase changes, a cause of fading of the signal. Related phenomena are the scintillation of distant radio sources due to the solar wind [6.4], and the twinkling of starlight brought about by atmospheric fluctuations [6.5]. In the swimming pool effect [6.6], at both deep and shallow ends, the light on the floor of the pool is uniform, but in the center, rippling bands of light and shadow are observed. Examples of considerable interest are furnished by moving diffuse surfaces such as ground glass [6.7-9], and the dynamic scattering [6.10] exhibited by a thin layer of nematic liquid crystal under the influence of an applied electric field [6.11]. A number of authors [6.12-17] employ the notion of a RPS in connection with the propagation through turbulence. For strong turbulence confined to a thin atmospheric layer, relations between the scintillation spectrum and the spectrum of the refractive index were established. Spatio-angular correlation measurements of stellar-light scintillation produced by double stars were used to estimate the turbulence strength in turbulent layers [6.18].

It has been noted on several occasions that a rough surface behaves as a phase screen [6.19]. When a highly coherent light beam is reflected or transmitted by an optically rough surface, random intensity or speckle pattern is produced. By virtue of the central limit theorem, the statistics of the scattered field will be Gaussian if the field results from a coherent superposition of many independent randomly phased contributions. In this case, the two lowest-order correlation functions determine all the field correlations of higher order. For a surface with a single scale of roughness, the spatial correlation function is essentially proportional to the Fourier transform of the illuminated area projected along the line of sight [6.20]. For this reason, the spatial coherence characteristics of speckle patterns do not convey significant information in the Gaussian-field limit. This may be contrasted with the time evolution of such patterns which continues to provide useful information on scatterer motion [6.21]. It thus becomes evident that the conditions of non-Gaussian statistics should lead to the most interesting experimental results regarding the RPS properties. As indicated in [6.21], there are three situations for which the scattered field is not Gaussian-distributed.

I) Due to partially coherent or partially polarized illumination, the various contributions to the field are added incoherently, on an intensity basis [6.22].

II) The contributions are not randomly phased, e.g., in the case of a partially developed speckle pattern, formed if the roughness of the object is too small to give extinction of the specular field component [6.23].

III) The total number of contributions is small and the application of the central limit theorem breaks down [6.19,24].

The purpose of this chapter is to review the information on the RPS obtainable from the far-field data. (The significance of the Fresnel region is mentioned in Sects.6.3.2 and 6.3.3.) Particular emphasis will be laid on higher-order statistical properties of the scattered field [6.25].

On the basis of a discrete micro-area model of the RPS, introduced in Section 6.2.1, we develop a general description of the statistical properties of the scattered field in terms of the characteristic functional. Implicitly, it contains information about the field statistics of all orders. In the context of the RPS statistics, a classic problem of a nonlinear transformation of the Gaussian process is taken up in Section 6.2. Section 6.3 is devoted to investigation of the field and intensity (irradiance) correlations. In Section 6.4, we discuss enhanced fluctuations generated by a small number of scatterers.

Some explanation of the notation may be helpful. We shall use the symbol $\Gamma$ to designate a spatio-temporal correlation function, and the symbol W to designate a spatial correlation. For example, $\Gamma^{(2)}(x_1,x_2;x_2,x_1)$ denotes the second-order intensity correlation function, cf. (6.28), and $W(\underline{r}_1,\underline{r}_2)$ refers to the first-order spatial correlation. The spatial coordinates in the observation plane will usually be labeled by $\underline{r}$, while the coordinates in the plane of the screen by $\underline{\rho}$. In addition, the input field and input correlation functions in the plane of the screen will be distinguished by subscript 0, e.g., $u_0(\underline{\rho})$, $W_0(\underline{\rho}_1,\underline{\rho}_2)$.

## 6.1 Basic Formulation of the Statistical Problem

We shall outline two approaches to the problem of determining the statistics of light scattered by a RPS: the continuum approach, which is appropriate for analyzing low-order statistical properties of scattered light, and the micro-area approach, based on representation of the screen as a collection of individual scatterers. Within the framework of the latter model, we derive the characteristic functional of the scattered field which embodies the complete statistical information. When the number of scatterers is large, the Gaussian limit is shown to hold asymptotically.

## 6.1.1 Physical Models

Our basic assumption is that the effect of the diffusing screen on an electromagnetic field can be expressed in terms of a number of macroscopic constants. An attempt to describe the scattering from a rough surface in a manner related to the atomic constitution of matter was made by AGARWAL [6.26], who used the perturbative treatment of the Ewald-Oseen extinction theorem. Other authors [6.27-29] examined the diffraction of electromagnetic waves from rough surfaces and slightly rough thin films, taking the surface plasmon formation into account. The theory developed in [6.29] seems to be well adapted to the scattering of light from metal surfaces and the excitation of surface plasmons.

On a macroscopic level, adopting the continuum approach, one starts with the scalar diffraction theory modified in an appropriate way to approximate a scattering screen. This procedure, used in the study of diffracted radio waves [6.30-32], was subsequently extended to the granularity analysis of scattered laser light [6.33-34].

To formulate the basic equations, we apply the Rayleigh-Sommerfeld diffraction integral. With reference to Fig.6.1, the complex scalar amplitude corresponding to frequency $\omega = ck$, c denoting the velocity of light, at the observation point $\underline{r}$, is given as

$$u(\underline{r}) = \frac{k}{2\pi i r}\int_{z=0} u_0(\rho)\, e^{ik|\underline{r}-\underline{\rho}|}\, d^2\rho \quad . \tag{6.1}$$

In the Fresnel approximation, we write (6.1) in the form

$$u(\underline{r}) = \frac{ke^{ikr}}{2\pi i r}\int_{z=0} u_0(\rho)\, \exp\left\{-i\left[\underline{k}_\perp\cdot\underline{\rho}+k_z\zeta(\underline{\rho})\right]+ik\rho^2/2r\right\}d^2\rho \quad , \tag{6.2}$$

where $\underline{k} = k\underline{r}/r$, $\underline{k}_\perp = k_x\hat{x}+k_y\hat{y}$, and $\phi(\underline{\rho}) = k_z\zeta(\underline{\rho})$ is the phase fluctuation introduced by the fluctuating height coordinate $z = \zeta(\underline{\rho})$. Equation (6.2) enables one to investigate the non-Gaussian effects in the Fresnel zone. We discuss some of them in Section 6.3.

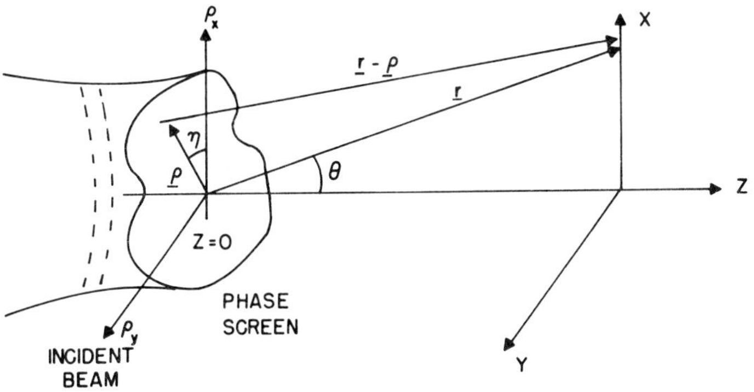


Fig. 6.1. Scattering geometry: Continuum model

If the illuminated area is much smaller than r/k, the Fraunhofer limit is valid and (6.2) reduces to

$$u(\underline{r}) = \frac{ke^{ikr}}{2\pi ir} \int_{z=0} u_0(\underline{\rho}) \exp\left\{-i\left[\underline{k}_\perp\cdot\underline{\rho}+k_z\zeta(\underline{\rho})\right]\right\}d^2\rho \quad . \tag{6.3}$$

Equation (6.3) restates BECKMANN's theory [6.35,36] in a form suitable for scattering in the optical region [6.37]. For a reflecting geometry, where $\underline{k}$ is to be interpreted as the change in the wave vector, a more rigorous justification of (6.3) was given by PEDERSEN [6.38], who started with Kirchhoff's diffraction integral.

Alternatively, the continuum model can be formulated in the language of scalar coherence theory [6.39], where only observable quantities appear. Having in mind the applications to image-forming systems, we introduce $K(\underline{r}-\underline{r}')$, the coherent point spread function in the image plane, corresponding to the point $\underline{r}'=\underline{\rho}$ in the object plane. The mutual spectral density, corresponding to a given realization of the transmittance $T(\underline{\rho})=\exp[-ik\zeta(\underline{\rho})]$, is

$$W(\underline{r}_1,\underline{r}_2) = \int_{z=0}\int_{z=0} W_0(\underline{\rho}_1,\underline{\rho}_2)T^*(\underline{\rho}_1)T(\underline{\rho}_2)K^*(\underline{r}_1-\underline{\rho}_1)\cdot K(\underline{r}_2-\underline{\rho}_2)d^2\rho_1\, d^2\rho_2 \quad . \tag{6.4}$$

In the case of a moving scattering surface, the transmittance will, in addition, be time-dependent. We note that (6.4) includes the free-space propagation problem.

The foregoing equations constitute a starting point to calculate the two- and fourth-point amplitude correlations. Unfortunately, the evaluation of statistical properties of higher order becomes progressively more difficult. It is the micro-area model which is intended to remedy the situation. In the context of the RPS problem, this model was used by ENLOE [6.40], ESTES et al. [6.41] and JAKEMAN and PUSEY [6.19,6.24]. However, the basic ideas can be traced back to the works of RAYLEIGH [6.42,43] and VON LAUE [6.44-46] (cf. also the diffraction problem by a random distribution of openings [6.47,48]).

Let us consider an incident beam of light with a well-defined direction of propagation $\hat{m}_0$,

$$u^{(inc)}(\underline{r},\omega) = u_0(\underline{r},\omega)\exp(i\underline{k}_0\cdot\underline{r}) \quad . \tag{6.5}$$

Here $\underline{k}_0=(\omega/c)\hat{m}_0$, and the amplitude function $u_0(\underline{r},\omega)$ will typically describe a laser beam Gaussian cross-section profile. Referring to Fig.6.2, we denote by $\underline{\rho}_j$ the location of the $j_{th}$ micro-area with respect to the origin. In the far zone, the contribution to the scattered field arising from that micro-area is then approximately

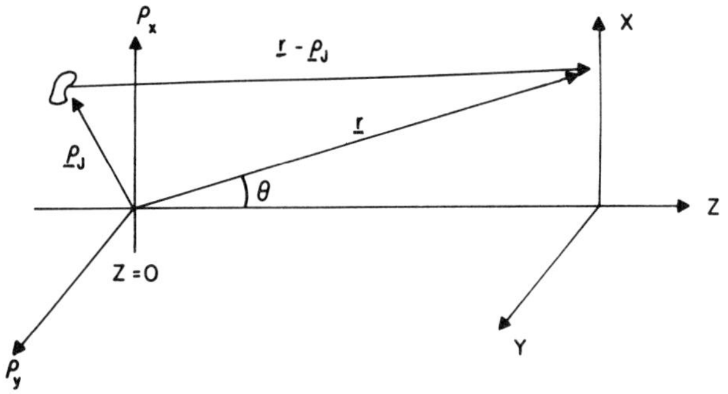


Fig. 6.2. Scattering geometry: Micro-area model

$$u_j(\underline{r},\omega) = u_0(\underline{\rho}_j,\omega)s_j \exp(i\alpha_j)\exp[i(kr-\underline{k}\cdot\underline{\rho}_j)] \quad , \tag{6.6}$$

where the scattering vector $\underline{k}=\underline{k}_s-\underline{k}_0$, with $\underline{k}_s=(\omega/c)\underline{r}/r$. The $\alpha_j$ denotes a random phase, while $s_j$, a real shape factor that accounts for diffraction produced by the micro-area around $\underline{\rho}_j$, is given by

$$s_j^2(\theta) = \left(\frac{k}{2\pi r}\right)^2 \iint\limits_{R_j} \exp\left[ik|\underline{\rho}_1-\underline{\rho}_2|\sin\theta\ \cos\eta+\phi_j(\underline{\rho}_1)-\phi_j(\underline{\rho}_2)\right]d^2\rho_1\ d^2\rho_2 \quad , \tag{6.7}$$

where $\eta$ is the azimuthal angle of the vector $\underline{\rho}_1-\underline{\rho}_2$. Within the region $R_j$ of the $j_{th}$ the micro-area, the phase $\phi_j$ of the wave front emerging from the phase screen changes coherently. This region is assumed to be sufficiently small so that, over the extent of $R_j$, the incident field is approximately constant. In general, both $\alpha_j$ and $\phi_j(\underline{\rho})$ are time-dependent stochastic functions.

The total scattered field, being the result of the coherent superposition of elementary contributions given by (6.6), is

$$u(\underline{r},\omega) = e^{ikr}\sum_j u_0(\underline{\rho}_j,\omega)s_j \exp(i\alpha_j)\exp(-i\underline{k}\cdot\underline{\rho}_j) \quad . \tag{6.8}$$

When the scattering screen is set in motion, with each micro-area moving along a trajectory $\underline{\rho}_j(t)$, an approximate expression for $V(\underline{r},t)$, the time-dependent optical disturbance, is

$$V(\underline{r},t) = \exp[i(kr-\omega_0 t)]\sum_j V_0[\underline{\rho}_j(t),t]s_j \exp(i\alpha_j)\exp[-i\underline{k}\cdot\underline{\rho}_j(t)] \quad . \tag{6.9}$$

This expression is valid for a quasi-monochromatic incident field characterized by the midfrequency $\omega_0$, under the assumption that $|\dot{\rho}(t)| \ll c$ [6.49,50].

According to the introduced terminology, the diffuse surface model used by GOLDFISHER [6.51], containing an infinitely dense collection of scatterers, can be regarded as pertaining to both our models. More refined models of the scattering surface are now available [6.52-57]. CROCE and PROD'HOMME [6.58] approached the problem by a vectorial method applying the Lorentz reciprocity principle.

### 6.1.2 Characteristic Functional of the Scattered Light

Since optical fluctuations are analyzed experimentally by means of photocount distribution measurements, it is the probability distribution, P(I), of the intensity of scattered light that acquires the primary importance. From the standpoint of the coherence theory of light, P(I) describes only one aspect of the problem, as it characterizes the field at one space-time point. A more general statistical description is embodied in the hierarchy of correlation functions that are determined by the joint probability distribution, $p_L$, of the optical field at L space-time points. In the limit, $L\to\infty$, $p_L$ becomes a probability functional, $p[V(\underline{r},t)]$, characterizing the field statistics completely [6.59-64].

For practical reasons, rather than evaluate $p[V(\underline{r},t)]$, we shall concentrate on the characteristic functional, $\Phi$, of the field. This is defined as a functional Fourier transform of p[V], similarly to the characteristic function, $\phi_L$, the L-dimensional Fourier transform of $p_L$.

Our task is to derive the characteristic functional on the basis of the micro-area model of Section 6.1.1. To this end we suppose that the number of scatterers situated within the illuminated area, $A_0$, of the RPS constitutes a random Poisson process [6.65-67]

$$N(A_0) = \int_{A_0} dN(\underline{\rho}) \quad . \tag{6.10}$$

This means that apart from the interference fluctuations generated by random $s_j$ and $\alpha_j$, we also account for the so-called occupation number fluctuations. The independent increments $dN(\underline{\rho})$ represent the random number of scatterers within $d^2\rho$ around $\underline{\rho}$. They are distributed with the average density $n(\underline{\rho})$ which determines the characteristic function of $dN(\underline{\rho})$:

$$\langle\exp[\mu dN(\underline{\rho})]\rangle = \exp\{n(\underline{\rho})[\exp(\mu)-1]d^2\rho\} \quad . \tag{6.11}$$

In the case where $n(\underline{\rho})$ is also a random function, one speaks about a compound or mixed Poisson process, otherwise the process is pure. For the mixed process, (6.11) is further averaged over the mixing distribution of n [6.68,69].

In terms of $dN(\underline{\rho})$, (6.9) is conveniently rewritten as

$$V(\underline{r},t) = \int R(\underline{r},t;\underline{\rho})dN(\underline{\rho}) \quad , \tag{6.12}$$

where

$$R(\underline{r},t;\underline{\rho}) = \exp[ikr-i\underline{k}\cdot\underline{\rho}(t)-i\omega_0 t]V_0[\underline{\rho}(t),t]s(\underline{\rho})\exp[i\alpha(\underline{\rho})] \quad . \tag{6.13}$$

Equation (6.12) defines $V(\underline{r},t)$ as a random function derived from the Poisson process [6.70]; the integration in (6.12) is to be interpreted in the Lebesgue-Stieltjes sense. In this way, both discrete and continuous distributions of scatterers are treated formally on an equal footing.

The characteristic functional of the scattered field is defined as

$$\Phi[\zeta(\underline{r},t)] = \langle\exp[\int\zeta(\underline{r},t)V^*(\underline{r},t)d^3rdt - \int\zeta^*(\underline{r},t)V(\underline{r},t)d^3rdt]\rangle \quad , \tag{6.14}$$

where the angular brackets indicate an ensemble average with respect to all the random variables in (6.12). For N scattering centers, N random phases $\alpha_j$ and N random positions $\underline{\rho}_j$ form together with N itself a 2N+1 dimensional random vector [6.40,71]. In addition, averaging over the distribution of shape factors $s_j$, and, for the mixed Poisson process, over $n(\underline{\rho})$, may be required. Where confusion may arise, the various averages will be distinguished by an appropriate subscript at the angular brackets.

Introducing the abbreviation

$$H(\underline{\rho}) = \int[\zeta(x)R^*(x,\underline{\rho})-\zeta^*(x)R(x,\underline{\rho})]d^4x \quad , \tag{6.15}$$

where we have used a shorthand notation $x = (\underline{r},t)$, $d^4x = d^3rdt$, the characteristic functional is written as

$$\Phi[\zeta] = \langle\exp\int H(\underline{\rho})dN(\underline{\rho})\rangle_{\alpha,\underline{\rho},s,N} \quad . \tag{6.16}$$

By virtue of the statistical independence of the increments $dN(\underline{\rho})$, this yields

$$\psi = \ln\Phi = \int d\psi(\underline{\rho}) \quad , \tag{6.17}$$

where

$$d\psi(\underline{\rho}) = \ln\langle\exp H(\underline{\rho})dN(\underline{\rho})\rangle_{\alpha,\underline{\rho},s,N} \quad . \tag{6.18}$$

From (6.11), we now obtain for infinitesimally small surface element $d^2\rho$ [6.72]

$$\begin{aligned} d\psi(\underline{\rho}) &= \ln\left\langle\exp\left(n(\underline{\rho})\{\exp[H(\underline{\rho})]-1\}d^2\rho\right)\right\rangle_{\alpha,\underline{\rho},s} \\ &= n(\underline{\rho})\langle\exp[H(\underline{\rho})]-1\rangle_{\alpha,\underline{\rho},s}\, d^2\rho \quad . \end{aligned} \tag{6.19}$$

Therefore

$$\Phi[\zeta] = \exp\left\{\bar{N} \int_{z=0} \langle \exp[H(\underline{\rho})]-1\rangle_{\alpha,s}\, n_0(\underline{\rho})d^2\rho\right\} \quad , \tag{6.20}$$

where we have factored out the average number of scatterers, $\bar{N}$, from $n(\underline{\rho})$. The resulting density $n_0(\underline{\rho})$, normalized to unity, can be considered as a probability density of location of a micro-area at $\underline{\rho}$.

In the case of a mixed Poisson process, (6.20) is further averaged over the ensemble of $n_0(\underline{\rho})$, giving

$$\Phi[\zeta] = \left\langle \exp\left\{\bar{N} \int_{z=0} \langle \exp[H(\underline{\rho})-1]\rangle_{\alpha,s}\, n_0(\underline{\rho})d^2\rho\right\}\right\rangle_{n_0} \quad . \tag{6.21}$$

By functional differentiation of the characteristic functional, the spatio-temporal coherence functions of the scattered field are obtained. Equation (6.20) or (6.21) for random $n_0$ generalizes therefore the results of [6.66], where a characteristic function appropriate for detection at one spatial point was derived.

### 6.1.3 Correlation Functions

By virtue of the defining equation (6.14), and from the definition of a functional derivative, the (n+m)th-order field amplitude moment involves the (n+m)th-order functional differentiation, viz.,

$$\langle V^*(x_1)\ldots V^*(x_n)V(x_{n+1})\ldots V(x_{n+m})\rangle$$

$$= \left.\frac{\delta^{n+m}\,\Phi}{\delta\zeta(x_1)\ldots\delta\zeta(x_n)\delta[-\zeta^*(x_{n+1})]\ldots\delta[-\zeta^*(x_{n+m})]}\right|_{\zeta=0} \quad . \tag{6.22}$$

In particular, the average amplitudes are

$$\langle V^*(x)\rangle = \bar{N}\left\langle \int \langle R^*(x,\underline{\rho})\rangle_{\alpha,s}\, n_0(\underline{\rho})d^2\rho\right\rangle_{n_0} \tag{6.23}$$

and

$$\langle V(x)\rangle = \bar{N}\left\langle \int \langle R(x,\underline{\rho})\rangle_{\alpha,s}\, n_0(\underline{\rho})d^2\rho\right\rangle_{n_0} \quad . \tag{6.24}$$

These functions, describing the mean scattered field, will be different from zero when the RPS is weak, i.e., producing phase variations $\alpha$ of less than $2\pi$. For a deep phase screen, the $\alpha$-variations are much greater than $2\pi$ and both $\langle R^*\rangle$ and $\langle R\rangle$ average out to zero.

The mutual coherence function is

$$\Gamma(x_1,x_2) = \langle V^*(x_1)V(x_2)\rangle = \overline{N}\langle\int\langle R^*(x_1,\underline{\rho})R(x_2,\underline{\rho})\rangle_{\alpha,s}\, n_0(\underline{\rho})d^2\rho\rangle_{n_0}$$
$$+\overline{N}^2\langle\int\langle R^*(x_1,\underline{\rho}_1)\rangle_{\alpha,s}\, n_0(\underline{\rho}_1)d^2\rho_1\int\langle R(x_2,\underline{\rho}_2)\rangle_{\alpha,s}\, n_0(\underline{\rho}_2)d^2\rho_2\rangle_{n_0} \quad . \tag{6.25}$$

In the remainder of this chapter, we shall consider only the case of nonrandom density of scatterers, so that the average over $n_0$ will be dropped.

The deep phase screen being characterized by the condition

$$\langle[R^*(x_1,\underline{\rho})]^n[R(x_2,\underline{\rho})]^m\rangle_{\alpha,s} = \delta_{nm}\langle[R^*(x_1,\underline{\rho})]^n[R(x_2,\underline{\rho})]^m\rangle_s \quad , \tag{6.26}$$

Equation (6.25) simplifies to

$$\Gamma(x_1,x_2) = \overline{N}\int\langle R^*(x_1,\underline{\rho})R(x_2,\underline{\rho})\rangle_s\, n_0(\underline{\rho})d^2\rho \quad . \tag{6.27}$$

The intensity correlation function or the second-order field correlation, at $x_1$ and $x_2$ is given as

$$\Gamma^{(2)}(x_1,x_2;x_2,x_1) = \langle I(x_1)I(x_2)\rangle = \left.\frac{\delta^4\Phi}{\delta\zeta(x_1)\delta\zeta(x_2)\delta[-\zeta^*(x_2)]\delta[-\zeta^*(x_1)]}\right|_{\zeta=0} \quad . \tag{6.28}$$

This expression, in general, contains 15 terms. For a deep phase screen, by virtue of (6.26), only three terms in (6.28) will survive, and the intensity correlation reduces to

$$\langle I(x_1)I(x_2)\rangle = \overline{N}\int\langle|R(x_1,\underline{\rho})|^2|R(x_2,\underline{\rho})|^2\rangle_s\, n_0(\underline{\rho})d^2\rho$$
$$+\overline{N}^2\left|\int\langle R^*(x_1,\underline{\rho})R(x_2,\underline{\rho})\rangle_s\, n_0(\underline{\rho})d^2\rho\right|^2$$
$$+\overline{N}^2\int\langle|R(x_1,\underline{\rho}_1)|^2\rangle_s\, n_0(\underline{\rho}_1)d^2\rho_1\int\langle|R(x_2,\underline{\rho}_2)|^2\rangle_s\, n_0(\underline{\rho}_2)d^2\rho_2 \quad . \tag{6.29}$$

It is readily seen that the first term in (6.29) describes a correction to Gaussian statistics, significant for small values of $\overline{N}$.

For the sake of simplicity, we write the general expression (6.28) in the special case of a reflective phase screen where, for specular direction, we can, in (6.13), set $\underline{k} = 0$. In addition, to simplify further, we neglect the variations of $V_0$, s, and $\alpha$ with $\underline{\rho}$, setting $V_0 = s = 1$. From (6.28), we then obtain

$$<I_1 I_2> = \bar{N}^4 |<R_1>|^2 |<R_2>|^2 + \bar{N}^3 \Big[ (<R_1^* R_2><R_1><R_2^*>+c.c) + (<R_1 R_2><R_1^*><R_2^*>+c.c) + <|R_1|^2>|<R_2>|^2 + |<R_1>|^2 <|R_2|^2> \Big] + \bar{N}^2 \Big[ (<R_1><R_1^*|R_2|^2>+c.c) + (<|R_1|^2 R_2^*><R_2>+c.c) + <R_1^* R_2><R_2^* R_1> + <R_1^* R_2^*><R_1 R_2> + <|R_1|^2><|R_2|^2> \Big] + \bar{N}<|R_1|^2 |R_2|^2> \quad , \qquad (6.30)$$

where the subscripts 1 and 2 refer to points $x_1$ and $x_2$, respectively, and the averaging operation is over the random distribution of $\alpha$. In a more general case, one would have to reintroduce the integrals over $\underline{\rho}$, possibly weighted with $n_0(\underline{\rho})$.

### 6.1.4 Gaussian Limit

When the average number of micro-areas is very large, by virtue of the central limit theorem the statistics of the scattered light becomes Gaussian. We can demonstrate this statement by investigating an asymptotic limit of the characteristic functional of the reduced variable $v(x) = \bar{N}^{-1/2}[V(x) - <V(x)>]$. In fact, using (6.20) we get

$$\Phi_{\bar{N}}[\zeta] = \exp\Big(\bar{N} \int d^2\rho\, n_0(\underline{\rho}) \Big\{ \bar{N}^{-1/2} <[H(\underline{\rho}) - <H(\underline{\rho})>]>_{\alpha,s} + \frac{1}{2\bar{N}} <[H(\underline{\rho})]^2>_{\alpha,s} + O(\bar{N}^{-3/2}) \Big\} \Big) \quad . \qquad (6.31)$$

As $\bar{N} \to \infty$, (6.31) takes the asymptotic form

$$\Phi_{\bar{N}\to\infty}[\zeta] = \exp\Big[ -\iint \zeta(x_1) \tilde{\Gamma}^{(1,1)}(x_1,x_2) \zeta^*(x_2) dx_1\, dx_2 + \frac{1}{2} \iint \zeta(x_1) \tilde{\Gamma}^{(2,0)}(x_1,x_2) \zeta(x_1) dx_1\, dx_2 + \frac{1}{2} \iint \zeta^*(x_1) \tilde{\Gamma}^{(0,2)}(x_1,x_2) \zeta^*(x_1) dx_1\, dx_2 \Big] \quad , \qquad (6.32)$$

where $G^{(1,1)}$ is given by (6.27) divided by $\bar{N}$, while $\tilde{\Gamma}^{(2,0)}$ and $\tilde{\Gamma}^{(0,2)}$ are given by similar equations involving, respectively, only $R^*$ and $R$. Equation (6.32) represents the characteristic functional of a noncircular Gaussian process with zero mean. When $\tilde{\Gamma}^{(2,0)}$ and $\tilde{\Gamma}^{(0,2)}$ vanish, one recovers the characteristic functional representing circular statistics of thermal light [6.73,74]. This will be the case when condition expressed by (6.26) is satisfied, i.e., when the incident beam is scattered by a deep RPS.

## 6.2 More General Detection and Coherence Conditions

In this section, we depart from a somewhat idealized situation considered thus far of detection at discrete space-time points. Instead, we shall account for a finite size of the scanning aperture, as well as for the finite response time of the detecting device, in the case of the Gaussian, possibly partially polarized, scattered light.

It is important to realize that there are two distinct physical situations in which one encounters sums of speckle patterns resulting from scattering by a phase screen. First, the light incident on the screen may be coherent, but one is interested in the statistics of the pattern observed through a scanning aperture with a finite response time. Second, the light incident on the screen is partially coherent, and the properties of the scattered light are analyzed by subdividing the resulting intensity into contributions arising from individual coherent components. As far as the spatial coherence is concerned, similarities of both situations have been stressed in recent reviews of the speckle statistics [6.22,75]. We consider, to begin with, the scattered field to be a Gaussian random process. Then we discuss a more realistic situation in which, due to a familiar convolution operation [6.76], the scattered field is not Gaussian, even though both the incident field and the fluctuating transmittance are Gaussian random processes. This will take place for polychromatic speckle patterns whose Fourier transformed amplitudes involve products of Gaussian variates.

### 6.2.1 Gaussian Scattered Field

In this subsection, we restrict our analysis to a deep RPS with a large mean number of scatterers, giving rise to a scattered light characterized by the Gaussian statistics. We briefly discuss the effects of the detection time and of the finite size of the scanning aperture in the general case of arbitrary polarization. Our analysis follows closely the theory developed earlier in connection with photoelectron counting [6.77].

If the aperture is considered to move cyclically with periodicity T, equal to the response time of the detecting device or to its multiplicity [6.78], then the observed intensity $I_A$ can be expressed in the form [6.79]

$$I_A = \frac{1}{TS}\int d^2r A(\underline{r}) \int_{t-T/2}^{t+T/2} I(\underline{r},t)dt \quad , \tag{6.33}$$

where $A(\underline{r})$ is a real and positive weighting function such that

$$\int A(\underline{r})\, d^2r = S \tag{6.34}$$

and S is the area of the aperture.

In (6.33), the intensity $I(\underline{r},t)$ is assumed to correspond to a plane wave with an arbitrary state of polarization. We thus write

$$I(\underline{r},t) = \sum_{\mu} V_{\mu}^{*}(\underline{r},t)V_{\mu}(\underline{r},t) \qquad (6.35)$$

where the Greek subscripts refer to rectangular x, y-components of the analytic signal $\underline{V}(\underline{r},t)$.

The statistical properties of $I_A$ are conveniently described in terms of the characteristic function

$$G(\xi) = \exp(i\xi I_A) \qquad (6.36)$$

which is given in the form of an infinite product

$$G(\xi) = \prod_{m} [1-(i\xi/TS)\kappa_m]^{-1} \qquad (6.37)$$

involving the eigenvalues, $\kappa_m$, of the matrix integral equation

$$\sum_{\nu} \int d^2r_2 A(\underline{r}_2) \int_{t-T/2}^{t+T/2} \Gamma_{\mu\nu}^{*}(\underline{r}_1,t_1,\underline{r}_2,t_2) f_{\nu}^{(m)}(\underline{r}_2,t_2)dt_2 = \kappa_m f_{\nu}^{(m)}(\underline{r}_1,t_1) \quad . \qquad (6.38)$$

The coherence matrix, forming the kernel of (6.38), is defined

$$\Gamma_{\mu\nu}(\underline{r}_1,t_1;\underline{r}_2,t_2) = \langle V_{\mu}^{*}(\underline{r}_1,t_1)V_{\nu}(\underline{r}_2,t_2)\rangle \quad . \qquad (6.39)$$

If (6.39) is effectively known, $G(\xi)$ can be approximated by the formula

$$G(\xi) = [1-(i\xi/TS)\kappa]^{-M} \quad , \qquad (6.40)$$

where we have assumed that the integral equation (6.38) has M equal eigenvalues $\kappa$. The requirement that the first two moments of the distribution $P(I_A)$ be exact leads to

$$M = TS/\Xi(T,S) \qquad (6.41)$$

and

$$\kappa = \langle I_A\rangle ST/N \quad , \qquad (6.42)$$

where

$$\Xi(T,S) = \frac{\sum_{\mu}\int d^2r_1 A(\underline{r}_1)\int d^2r_2 A(\underline{r}_2)\iint_{t-T/2}^{t+T/2} |\Gamma_{\mu\nu}(\underline{r}_1,t_1;\underline{r}_2,t_2)|^2\, dt_1\, dt_2}{ST\langle I_A\rangle^2} \tag{6.43}$$

and

$$\langle I_A\rangle = \frac{1}{TS}\sum_{\mu}\int d^2r\; A(\underline{r})\int_{t-T/2}^{t+T/2} \Gamma_{\mu\nu}(\underline{r},t;\underline{r},t)dt \quad . \tag{6.44}$$

By the Fourier inversion formula applied to (6.40), and with the help of (6.42), we get

$$P(I_A) = \frac{M^M}{\Gamma(M)}\frac{I_A^{M-1}}{\langle I_A\rangle^M}\exp(-MI_A/\langle I_A\rangle) \quad , \tag{6.45}$$

which is a gamma density distribution, with the parameter M specifying the number of the degrees of freedom of light [6.80-92]. Asymptotically, as T and S tend to infinity, $\Xi(T,S)$ is to be identified with the volume of coherence [6.93], and M can then be interpreted as the number of correlation cells. We emphasize, however, that otherwise numbers of degrees of freedom and of correlation cells are different [6.79, 82]. In any event, M is always related to the contrast $C_A$ of $I_A$, defined as the ratio of the standard deviation to the mean value, through the equation

$$C_A = M^{-1/2} \quad . \tag{6.46}$$

As an illustration of (6.46) in the time domain, we show in Fig.6.3 three photographs of the speckle patterns produced by a rotating ground glass, corresponding to three different ratios of the observation (exposition) time, T, to the coherence time. Since the number of degrees of freedom increases with T, $C_A$ decreases, in accordance with (6.46).

When the speckle pattern is recorded by a logarithmic device, such as a photographic film, the variate $I_A$ is transformed logarithmically. This was discussed by BARAKAT [6.94], and by ARSENAULT and APRIL [6.95].

In a number of situations of practical interest, the light is both spectrally [6.96-98] and polarizationally [6.77] pure, leading to a factorized form of $\Gamma(\underline{r}_1,t_1,\underline{r}_2,t_2)$. Specifically, let us suppose that the following reduction formula holds

$$\Gamma_{\mu\nu}(\underline{r}_1,t_1;\underline{r}_2,t_2) = (1/2)\gamma_{\mu\nu}(\underline{r}_0,\underline{r}_0)W(\underline{r}_1,\underline{r}_2)\gamma(t_1-t_2) \quad , \tag{6.47}$$

where $\gamma_{\mu\nu}(\underline{r}_0,\underline{r}_0)$, the normalized coherence matrix at a reference point $\underline{r}_0$, describes the state of polarization, while $\gamma(t_1-t_2)$ is the normalized complex degree of tem-

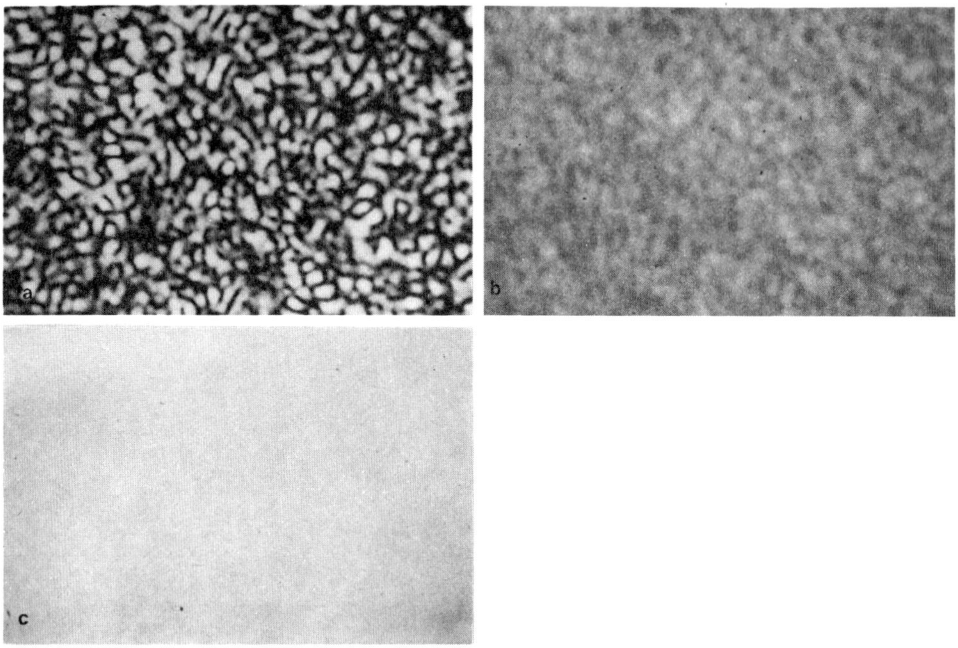

Fig. 6.3a-c. Speckle patterns produced by a pseudothermal source. The source was a rotating ground glass with scattering grains having the diameter of the order of 10μ, illuminated by a He-Ne laser beam, $\lambda = 6328$ Å. At the circular aperture, with diameter 0.35 mm, which diaphragmed the ground glass, the linear speed of rotation was $v = 1$ mm/s. The coherence time, corresponding to interference fluctuations, was estimated to be $T_C = \lambda/v = 0.6$ ms. a) Diffuser at rest $T_C = \infty$, $M = 1$, $C_A = 1$; b) exposition time $T = 1/100$ s., $M = T/T_C = 16.7$, $C_A = 0.2$; c) $T = 1$s., $M = 1666.7$, $C_A = 0.02$ (Courtesy of J. Bures)

poral coherence. The factor 1/2 accounts for the fact that, with an appropriate choice of axes, the average intensity associated with x or y components of the field is just one half of the total average intensity.

If (6.47) is satisfied, (6.37) takes the form

$$G(\xi) = \prod_{m,n} \left[1-(i\xi/2TS)\,(1-P)\lambda_m\nu_n\right]^{-1}\left[1-(i\xi/2TS)(1+P)\lambda_m\nu_n\right]^{-1} \quad , \tag{6.48}$$

where $\lambda_m$ and $\nu_n$ are the eigenvalues of integral equations with kernels $W(\underline{r}_1,\underline{r}_2)$ and $\gamma(t_1-t_2)$, respectively. The approximate formulas (6.40) or (6.45) still hold, but the function $\Xi(T,S)$ now factorizes into the product of three factors

$$\Xi(T,S) = [(1+P^2)/2]\Xi_1(T)\Xi_2(S) \quad , \tag{6.49}$$

where the explicit form of $\Xi_1$ and $\Xi_2$ is easily deduced from (6.43) and (6.47). If $T_c$ and $S_c$ denote the coherence time and coherence area at the detector surface, then for $T \ll T_c$ and $S \ll S_c$, only polarizational degrees of freedom are significant. The exact form of the distribution $P(I_A)$ coincides with the probability density of a partially polarized light [6.79,84,99]. An approximate form of $P(I_A)$ is given by (6.45) with $M = 2/(1+P^2)$. Some other aspects of the polarization effects were investigated by SCRIBOT [6.100] and REZETTE [6.101]. Depolarization of light backscattered from rough surfaces is covered in [6.102-107].

If, on the other hand, $P = 1$ and $T \ll T_c$, one recovers the problem of the distribution of intensity measured with finite aperture [6.94,108-112], or, equivalently, the intensity produced by partially coherent light detected at one point. In relation to the inverse problem, we now establish a connection between the number of degrees of freedom, M, and the properties of the near field. To this end, we observe that, since the radiation field immediately behind the screen can be regarded as spatially incoherent, at least for sufficiently fine microstructure of the scattering screen surface, we can evaluate M from (6.41) and (6.43) with the help of the Van Cittert-Zernike theorem [6.48]. If $P = 1$ and $T \ll T_c$, after a familiar change of variables [6.113,114], letting $W(\underline{\varrho}_1,\underline{\varrho}_2) = I_0(\underline{\varrho}_1)\delta(\underline{\varrho}_1-\underline{\varrho}_2)$, we obtain

$$\Xi(T,S) = \frac{T\int\limits_{z=0} d^2\rho_1 I_0(\underline{\varrho}_1)\int\limits_{z=0} d^2\rho_2 I_0(\underline{\varrho}_2)\int d^2r' \exp\left[\frac{2\pi i}{D\lambda_0}\underline{r}'\cdot(\underline{\varrho}_1-\underline{\varrho}_2)\right]\int B\left(\underline{r}+\frac{\underline{r}'}{2}\right)B\left(\underline{r}-\frac{\underline{r}'}{2}\right)d^2r}{S\left[\int d^2\rho I_0(\underline{\varrho})\right]^2} , \tag{6.50}$$

where $\lambda_0$ denotes a mean wavelength, while D is the distance between the phase screen and the scanning aperture. In (6.50), $I_0(\underline{\varrho})$ denotes frequency-integrated intensity distribution. For $S \gg S_c$, (6.41) and (6.50) yield the desired result

$$M^{-1} = \frac{D^2\lambda_0^2}{S}\frac{\int\limits_{z=0}[I_0(\underline{\varrho})]^2\,d^2\rho}{\left[\int\limits_{z=0} I_0(\underline{\varrho})d^2\rho\right]^2} , \tag{6.51}$$

which tells us that the squared contrast of the observed speckle pattern is proportional to a normalized second moment of intensity immediately behind the phase screen, provided that $S \gg S_c$. A generalization of (6.51), accounting for a finite spectral width of the radiation, reads

$$M^{-1} = \frac{D^2\lambda_0^2}{ST}\frac{\int\limits_0^\infty d\omega\int\limits_{z=0}[I_0(\underline{\varrho},\omega)]^2\,d^2\rho}{\left[\int\limits_0^\infty d\omega\int\limits_{z=0} I(\underline{\varrho},\omega)d^2\rho\right]^2} . \tag{6.52}$$

When the light is cross-spectrally pure, and the spatial distribution of intensity behind the screen can be regarded as uniform (6.52) gives

$$M = (S/S_c)(T/T_c) \quad , \tag{6.53}$$

which coincides with our notion of M as a number of correlation cells.

### 6.2.2 Polychromatic Speckle Patterns

In a general case, (6.52) cannot rigorously be maintained, since the random features of the scattering screen cause the statistics of the scattered light to deviate from Gaussian. In fact, assuming spatially coherent polarized illumination of the scattering screen, by virtue of (6.3), we can write

$$u(\underline{r},t) = \int_0^\infty d\omega\; u^{(S)}(\omega)\; u^{(P)}(\underline{r},\omega)e^{-i\omega t} \quad , \tag{6.54}$$

where $u^{(S)}(\omega)$ describes the spectral content of the source, while

$$u^{(P)}(\underline{r},\omega) = \frac{ke^{ikr}}{2\pi ir}\int_{z=0} u_0(\underline{\rho})\; \exp\{-i[\underline{k}_\perp\cdot\underline{\rho}+k_z\zeta(\underline{\rho})]\}d^2\rho \tag{6.55}$$

is given in terms of the parameters of the phase screen. Suppressing for the moment the space dependence, and assuming both $u^{(S)}(\omega)$ and $u^{(P)}(\omega)$ to be Gaussian variates, we obtain for the first two moments of the observed intensity

$$\langle I_A\rangle = \int_0^\infty g(\omega)\langle|u^{(P)}(\omega)|^2\rangle d\omega \quad , \tag{6.56}$$

$$\begin{aligned}\langle I_A^2\rangle = &\int_0^\infty\int_0^\infty g(\omega_1)g(\omega_2)\Big[\langle|u^{(P)}(\omega_1)|^2\rangle\langle|u^{(P)}(\omega_2)|^2\rangle \\ &+ |\langle u^{(P)*}(\omega_1)u^{(P)}(\omega_2)\rangle|^2\Big]d\omega_1\, d\omega_2 \\ &+ \frac{1}{T^2}\int_{t_1-T/2}^{t_1+T/2} dt_1 \int_{t_2-T/2}^{t_2+T/2} dt_2 \iint g(\omega_1)g(\omega_2)\; \exp[i(\omega_1-\omega_2)(t_1-t_2)] \\ &\cdot \Big[\langle|u^{(P)}(\omega_1)|^2\rangle\langle|u^{(P)}(\omega_2)|^2\rangle + |\langle u^{(P)*}(\omega_1)u^{(P)}(\omega_2)\rangle|^2\Big]d\omega_1\, d\omega_2 \quad .\end{aligned} \tag{6.57}$$

Equation (6.56), with the spectral function $g(\omega) = \langle|u^{(S)}(\omega)|^2\rangle$ constitutes, together with (6.54), a starting point for studying the speckle patterns produced in polychromatic light [6.22,38,115-119]. Although (6.46) is no longer true, the speckle contrast can readily be found with the aid of (6.56) and (6.57) in the limit of very long and very short integration times [6.120]. For $T \gg T_c$, the second term in (6.57) is negligible, and we get

$$C_L = \frac{\int_0^\infty d\omega_1\, g(\omega_1) \int_0^\infty d\omega_2\, g(\omega_2) |\langle u^{(P)*}(\omega_1) u^{(P)}(\omega_2)\rangle|^2}{\int_0^\infty g(\omega) \langle |u^{(P)}(\omega)|^2 \rangle d\omega} \quad . \tag{6.58}$$

On the other hand, if $T \ll T_c$, then

$$C_S = 2C_L + 1 \quad . \tag{6.59}$$

Under the condition of narrow spectral bandwidth, centered around a midfrequency $\omega_0$, we can let $\langle u^{(P)*}(\omega_1)u^{(P)}(\omega_2)\rangle = \langle |u^{(P)}(\omega_0)|^2\rangle$ in (6.58) which, interestingly, expresses the contrast in terms of the temporal coherence function of the source.

Returning now to (6.57), we see that in the limit of a long integration time $\langle I_A^2 \rangle$ is a second-order moment of a Gaussian process with the spectral density playing a role of the weight function $A(\underline{r})$ for an extended detector. Therefore, in this limit, the theory outlined in Section 6.2.1 remains valid, if, instead of the time domain, the frequency domain is considered. In particular, (6.38) now becomes

$$\sum_\nu d^2 r_2 A(\underline{r}_2,\omega_2) \int_0^\infty \Gamma^*_{\mu\nu}(\underline{r}_1,\omega_1;\underline{r}_2,\omega_2)\, f_\nu^{(m)}(\underline{r}_2,\omega_2)\, d\omega_2 = \kappa_m\, f_\nu^{(m)}(\underline{r}_1,\omega_1) \quad , \tag{6.60}$$

which generalizes the theory of PARRY [6.22] to include the polarization effects.

## 6.3 Amplitude and Intensity Correlations

It is unlikely that the far-field data will ever make it possible to obtain a complete statistical description of a phase screen as a random process in two spatial dimensions. Instead, one has to content oneself with the statistical characteristics such as the variance or the autocorrelation of the surface profile. In the next two subsections, we discuss how the lowest-order ensemble averaged properties of the scattered field enable one to deduce the statistical features of the phase screen. In Section 6.3.2, we deal with the temporal coherence of light produced by a moving diffuser.

### 6.3.1 Information Contained in Amplitude Correlations

We start by evaluating the angular correlation function, using (6.3) in which we let $k_z \to k$. This replacement, which amounts to a simplification of the Beckmann-Pedersen theory, is discussed by WELFORD [6.121]. Introducing the spatial frequencies

$\underline{f}_i = \underline{k}_{\perp i}/2\pi = k\underline{s}_i/2\pi$, $i = 1,2$, and assuming Gaussian statistics for $\zeta(\underline{\rho})$, with the variance $\sigma^2$ and the autocorrelation coefficient $q(\underline{\rho})$, we obtain

$$W(\underline{s}_1,\underline{s}_2) = \left(\frac{k}{2\pi r}\right)^2 \exp[-ik(r_1-r_2)] \cdot \int\limits_{z=0} d^2\rho \; e^{-2\pi i k\underline{s}\cdot\underline{\rho}} \exp\{-k^2\sigma^2[1-q(\underline{\rho})]\}$$

$$\int\limits_{z=0} u_0^*\left(\underline{\rho}' + \frac{\underline{\rho}}{2}\right) u_0\left(\underline{\rho}' - \frac{\underline{\rho}}{2}\right) \cdot e^{-2\pi i k\underline{s}'\cdot\underline{\rho}'} \, d^2\rho' \quad , \tag{6.61}$$

where $\underline{s} = (\underline{s}_1+\underline{s}_2)/2$, $\underline{s}' = \underline{s}_1 - \underline{s}_2$. In (6.61), the input field $u_0(\underline{\rho})$ may be considered to account both for the inhomogeneity of the incident beam and for the aperture size.

Analytic results for the angular coherence and angular intensity implied by (6.61) were obtained by BALTES and STEINLE [6.122-124], for Gaussian intensity profile, $u_0(\underline{\rho}) = \exp(-\rho^2/4a^2)$, and Gaussian correlation coefficient (see also Sect.5.4.4). Their results, valid if $q(\underline{\rho})$ can be expanded in terms of the correlation length $\rho_c$, viz., $q(\underline{\rho}) \simeq 1 - \rho^2/\rho_c^2$, show how the knowledge of b, the effective correlation length of the phase screen, $b^2 = \rho_c^2/2k^2\sigma^2$, can be achieved from far-field measurements. For optimal results, one should measure the degree of angular coherence or the radiant intensity, according as $a \leqq b$ or $a \geqq b$. In the limit $b/a \ll 1$, no information about the scatterer is available from measuring the angular coherence. Some caution should be exercised, however, when expanding the autocorrelation coefficient, especially in the deep screen limit [6.125,126].

The work of LEADER [6.20], who formulated a vector theory of reflective scattering, contains a multiplicative correction to (6.61) in the form of an exponential decreasing with angular separation between $\underline{s}_1$ and $\underline{s}_2$. Leader then derives the spatial coherence function of light scattered from a rough surface described as a superposition of two random height distributions [6.53].

NAGATA and UMEBARA [6.127,128] determined both the variance and the correlation length of the height coordinate. Their starting point is an equation similar to (6.61), but modified to account for the wave front curvature of the incident laser beam. We also mention the work of TAGANOV and TOPORETS [6.129] who estimated the effect of the surface roughness on the degree of spatial coherence.

Broadly speaking, as shown by CHANDLEY and WELFORD [6.37,121,130,131], the light scattered into specular direction by a reflective phase screen gives information about the variance of the random phase, while that scattered away from the specular direction depends, in addition, on the autocorrelation of the phase shift. In order to see that, we use (6.61) to write for $\langle I(\underline{s})\rangle = W(\underline{s},\underline{s})$, the angular intensity

$$\langle I(\underline{s})\rangle = \left(\frac{k}{2\pi r}\right)^2 e^{-k^2\sigma^2} \Big|\int\limits_{z=0} u_0(\underline{\rho})\, e^{-2\pi i k\underline{s}\cdot\underline{\rho}} \, d^2\rho\Big|^2$$

$$+ F\int\limits_{z=0} e^{-k^2\sigma^2}\left(e^{k^2\sigma^2 q(\rho)} - 1\right) e^{-2\pi i k\underline{s}\cdot\underline{\rho}} \, d^2\rho \quad , \tag{6.62}$$

where the constant F comes from the integral over $\underline{\rho}'$ in (6.61) containing slowly varying input field distribution.

For the specular direction, where $\underline{s} = 0$, the first deterministic term, arising from $|\langle\exp(i\zeta)\rangle|^2$, is significant. The second term, important for nonspecular direction, is the Fourier transform of Q, the autocovariance of the complex random process $\exp(i\zeta)$. On inverting the transform, we obtain

$$Q(\rho) = \langle\exp[i(\zeta_1-\zeta_2)]\rangle - \langle\exp(i\zeta_1)\rangle\langle\exp(-i\zeta_2)\rangle$$

$$= \frac{1}{F}\int \langle I(\underline{f})\rangle\, e^{2\pi i \underline{f}\cdot\underline{\rho}}\, d^2f \quad . \tag{6.63}$$

To be sure, the inclination factor, F, depends on $\underline{f}$, but is frequently taken as a constant over the relevant range [6.131].

If we retain our assumption about the Gaussian statistics of the random height $\zeta$, then the measurement of $Q(\underline{\rho})$ yields the autocorrelation coefficient of $\zeta$ in the form

$$q(\rho) = 1 + (k\sigma)^{-2} \ln\left[\frac{Q(\rho)}{Q(0)}\left(1-e^{-k^2\sigma^2}\right) + e^{-k^2\sigma^2}\right] \quad . \tag{6.64}$$

It follows from the work of CHANDLEY [6.130] that the surfaces studied by him do not have a Gaussian autocorrelation coefficient. However, the intensity behavior does not depend critically on the Gaussian form of $Q(\rho)$, when the random screen is weak [6.132].

A well-known realization of the RPS studied intensively in recent years are nematic liquid crystals in the ordered phase [6.133-135] which scatter light under the application of an external electric field. The spatial and temporal first-order properties of laser light scattered by such a system were investigated by BARTOLINO et al. [6.136], and by SCUDIERI et al. [6.137]; related subjects are dealt with in [6.138-140]. BERTOLOTTI et al. [6.141] showed how the spatial coherence angle can be related to the fluctuations of the dielectric susceptibility constant of the medium, allowing for a determination of the correlation length of the fluctuations.

### 6.3.2 Information Contained in Intensity Correlations

A familiar manifestation of intensity fluctuations is the Wiener spectrum. This quantity, encountered mostly in the context of noise generation by speckling [6.40, 51,142-144], provides a suitable measure of coarseness or granularity of the speckle pattern. It is, however, the average contrast which rather is used for surface roughness estimation [6.23,38,118,145-159]. A comprehensive review of techniques for measuring surface roughness and optical figure was published by BENNET [6.160]. RICHMOND and HSIA [6.161] collected a bibliography on scattering by reflection from surfaces of different types.

For a class of simple surface models, the basic ingredients of the theory are contained in (6.25) and (6.30). Letting the points $x_1$ and $x_2$ to coincide, we obtain the first two intensity moments at a single point. Hence, the contrast is given as

$$C_I = \sigma_I/\langle I\rangle \qquad (6.65)$$

with

$$\sigma_I^2 = \bar{N}^3\left[2\langle|R|^2\rangle|\langle R\rangle|^2+(\langle R^2\rangle\langle R^*\rangle^2+\mathrm{c.c})\right] + \bar{N}^2\left[2(\langle R\rangle\langle R^*|R|^2\rangle+\mathrm{c.c})+\langle|R|^2\rangle^2+\langle R^{*2}\rangle\langle R^2\rangle\right] + \bar{N}\langle|R|^4\rangle \qquad (6.66)$$

and

$$\langle I\rangle = \bar{N}\langle|R|^2\rangle + \bar{N}^2|\langle R\rangle|^2 \quad . \qquad (6.67)$$

When R is identified with $\exp(i\alpha)$, the above equations coincide with PEDERSEN's result [6.147] averaged over the Poisson distribution of N, the number of scatterers. In Fig.6.4, we plot the speckle contrast as function of $\sigma_\alpha$, the rms phase deviation, which is proportional to the rms surface roughness of the object. It is evident that the contrast saturates at a level greater than one, viz., $(1+1/\bar{N})^{1/2}$, and that for $\sigma_\alpha$ sufficiently small the contrast does not depend appreciably on the detailed form of the phase distribution. A strong dependence on the choice of model, found by

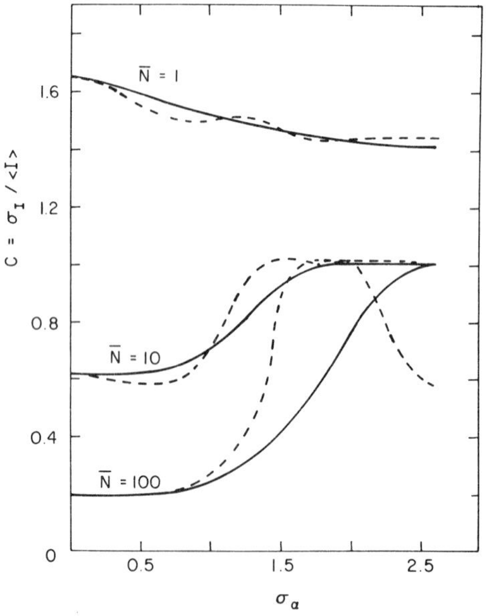


Fig. 6.4. Contrast of speckle patterns produced by a weak RPS. Solid curves correspond to Gaussian and dotted curves to rectangular phase distributions

PEDERSEN, is observed only in the range of large $\sigma_\alpha$, where the phase screen cannot be regarded as weak any longer, and the speckle becomes fully developed. It should also be evident that statistics of the scattered field is noncircular [6.147,162] whenever $\langle R\rangle$ and $\langle R^*\rangle$ do not vanish. A detailed investigation of the noncircular statistics in the Gaussian limit is to be found in the work of OHTSUBO and ASAKURA [6.163].

The intensity correlation properties in the image plane depend stronly on the mode of illumination [6.71,150,152], and on the properties of the optical system [6.157,164]. This seems to hinder the image-plane usefulness as compared to the far-field. Although recently derived formulas [6.145] have a complicated dependence on parameters of the optical system, an insight can be gained by combining the micro-area model developed in Section 6.1.1 with (6.4). If the complex transmittance is written in the form

$$T(\underline{\rho}) = \sum_{j=1}^{N} s_j \exp(i\alpha_j)\delta(\underline{\rho}-\underline{\rho}_j) \quad , \tag{6.68}$$

then, following the derivation of ICHIOKA [6.71], we obtain, after averaging over random $s_j$, $\alpha_j$, $\underline{\rho}_j$, and N, the expression for the contrast in the form

$$C_I^2 = \int |\tilde{K}(\underline{f})|^2 \left[\tilde{W}_0(\underline{f}) + \frac{2\langle s^4\rangle}{\langle s^2\rangle^2 \overline{n}}\right] d^2f \quad , \tag{6.69}$$

where $\overline{n}$ is the average number of scatterers per unit area, while

$$\tilde{K}_0(\underline{f}) = \frac{\int \hat{K}(\underline{f}+\underline{f}_1)\hat{K}^*(\underline{f}_1)d^2f_1}{\int |\hat{K}(\underline{f})|^2 \, d^2f} \tag{6.70}$$

and

$$\tilde{W}_0(f) = \frac{\int \hat{W}(\underline{f}+\underline{f}_1)\hat{W}^*(\underline{f}_1)d^2f_1}{\left[\int \hat{W}_0(\underline{f})d^2f\right]^2} \quad . \tag{6.71}$$

Here $\hat{K}(\underline{f})$ and $\hat{W}_0(\underline{f})$ denote the Fourier transforms of the point spread function, $K(\underline{\rho})$, and of the mutual spectral density, $W_0(\underline{\rho})$, respectively.

In the coherent limit, one can replace $\hat{W}(\underline{f})$ by a delta function. Then a rough qualitative estimate of (6.69) reads

$$C_I^2 = 1 + \frac{\langle\phi^2\rangle}{4\overline{N}} \tag{6.72}$$

where, to calculate $\langle s^{2n}\rangle$, we have used formulas derived on the basis of (6.7) by JAKEMAN and PUSEY [6.19,24], and then we have set the scattering angle $\theta = 0$. In a general case, a decrease of the spatial coherence of illuminating light reduces the contrast. However, as claimed by FUJII and ASAKURA [6.149,150], a linear relation between the contrast and the rms deviation of the surface roughness can, for less coherent illumination, be extended to a broader region. This conclusion, although subject to criticism [6.121], remains in agreement with an apparently different problem, studied by ROSS [6.166-168], of extracting structural information from light scattering by amorphous media.

Regarding the point spread function, it is inferred from the computer simulation data that it does not affect the contrast drastically [6.156,157,169].

It is noteworthy that (6.72), when remedied for the angle-dependent factor, describes correctly the speckle contrast, named also the scintillation index, in the far field. For random N, the previously mentioned result of JAKEMAN and PUSEY is

$$\sigma_I^2 = 1 + \frac{\langle\rho^2\rangle}{4\bar{N}} \exp\left(\frac{k^2\xi^2 \sin^2\theta}{4\langle\phi^2\rangle}\right) \tag{6.73}$$

under the approximate condition that $\phi$ varies linearly over the region of a micro-area which is of a dimension $\xi$, the correlation length. It is evident that from goniometric measurements, described in Chapter 5, see also [6.170-172], one can, in principle, infer all the structural parameters of interest, i.e., $\langle\phi^2\rangle$, $\xi$, and $\bar{N}$.

In the Fresnel region, the situation is complicated by the presence of focusing as well as overlapping and interference effects [6.173-176]. Both numerical and analytical integration of (6.2) shows that the intensity fluctuations increase from zero near the screen to a maximum in the region of focusing, the effect due to the lens-like behavior of individual refractive index inhomogeneities. A careful analysis of the diffraction integral enables one to identify "many-scatterer" terms responsible for speckle and "single-scatterer" terms describing focusing effects. The modeling of the phase correlation function, undertaken by JAKEMAN and McWHIRTER [6.175], should be pertinent to atmospheric propagation problems.

### 6.3.3 Moving Diffusers

The speckle patterns have increasingly been used to probe the diffusely reflecting bodies in order to extract information about their motion [6.177-193], strain [6.194,195], and mechanical oscillations [6.196,197]. Those applications are covered exhaustively by ENNOS [6.198]. Here we wish to establish a connection between the amplitude correlations of the scattered light and the velocity of a rotating phase screen. In this context, the intensity correlations, studied usually by photon correlation techniques [6.199], are also important.

Without entering into the analysis of other approaches [6.200-202], we can start directly with (6.25) in which, for a deep RPS, the second term is neglected. After invoking (6.13), in which the first exponential is supplemented with the term $ik\rho^2/2D$, describing the Fresnel-zone effects, in the stationary situation, we obtain for the mutual coherence function

$$\Gamma(\underline{r}_1,\underline{r}_2,\tau) = \bar{N}n_0 s^2 \exp\left[ik(r_1-r_2)-i\omega_0\tau+i\frac{k}{2D}(\underline{r}_1-\underline{r}_2)\cdot\underline{\nu}\tau\right]$$
$$\cdot\int_{z=0} V_0^*\left(\underline{\rho}-\frac{\underline{\nu}\tau}{2}\right)V_0\left(\underline{\rho}+\frac{\underline{\nu}\tau}{2}\right)\exp\left[i\frac{k}{D}(\underline{r}_1-\underline{r}_2)\cdot\underline{\rho}+i\frac{k}{2D}\left(\underline{\rho}-\frac{\underline{\nu}\tau}{2}\right)^2\right.$$
$$\left.+i\frac{k}{2D}\left(\underline{\rho}+\frac{\underline{\nu}\tau}{2}\right)^2\right]d^2\rho \quad . \qquad (6.74)$$

Here D is the distance between the RPS and the observation plane, and, for the sake of simplicity, we have taken $n_0$ and s to be independent of the location on the screen. Furthermore, at a sufficiently large distance from the rotation axis, the micro-areas will be in uniform motion with constant velocity $\underline{\nu}$. For an incident laser beam having the form $\exp(-\beta\rho^2)$, where the real part of the complex parameter $\beta$ describes the beam width, while the imaginary part describes the curvature of the wave front, (6.74) becomes

$$\Gamma(\underline{r}_1,\underline{r}_2,\tau) = Z(\underline{r}_1,\underline{r}_2)\cdot\exp\left\{-\frac{\mathrm{Re}\{\beta\}\nu^2\tau^2}{2}-\frac{\left[\left(\frac{k}{D}-2\mathrm{Im}\{\beta\}\right)\underline{\nu}\tau+\frac{k}{D}(\underline{r}_1-\underline{r}_2)\right]^2}{8\mathrm{Re}\{\beta\}}\right\} \quad , \qquad (6.75)$$

where $Z(\underline{r}_1,\underline{r}_2)$ is a complex amplitude whose explicit form is of no further importance for our considerations. In particular, when the laser beam with a waist w is focalized by means of a lens of focal length $f_\ell$, (6.75) yields the Gaussian temporal coherence function $\Gamma(\underline{r},\underline{r},\tau)$, which coincides with the result derived first by ESTES et al. [6.41]. Taking into account the surface parameters, such as the variance of the random phase or the correlation length, does not affect the form of the time correlation [6.203]. The effect of a finite scattering spot was studied in [6.66, 204]; the moving aperture method was also applied for laser speckle reduction [6.205].

The far-field power spectrum, investigated in [6.41], is a Gaussian function of frequency with the half-width $\Delta\omega_{1/2}$ directly proportional to the velocity of translation, viz., $\Delta\omega_{1/2} = (2\ln 2)^{1/2}\nu\cdot[(k^2w^2/4f_\ell^2)+w^{-2}]^{1/2}$. When the rotating object is small enough, one cannot approximate the velocity as linear; an analysis of a periodic random signal leads then to the determination of the rotation velocity [6.206].

As is revealed by (6.75), the dynamic behavior of the scattered field, due to the curvature of the incident wave front, or to the curvature of the wave front in the Fresnel-zone, shows both a bodily translation and a gradual change in structure. This was studied in greater detail both in the Fresnel region [6.207] and in the image field [6.208]. In the situation termed "boiling", predicted by (6.75) when $k/D = 2\mathrm{Im}\{\beta\}$, the velocity of the speckle translation vanishes, and the speckles merely change their structure with time. This leads to a host of practical applications, such as determination of pattern correlations, modulation transfer function of an imaging system, the accomodation state of the eye [6.209-211], and camera testing [6.212].

## 6.4 Number-Dependent Effects

When only a small number of scatterers contribute to the amplitude of the scattered field, deviations from the Gaussian-field limit will depend both on the number and properties of individual contributions. The results expressed by (6.72,73) show the significance of the contrast-enhancing second term in providing information about the parameters of the scattering screen. Equipped with the necessary formalism of Section 6.1, we are in position to treat in some detail the higher-order statistics of the non-Gaussian fluctuations generated by a deep RPS.

### 6.4.1 Moments and Probability Distribution of Intensity

The intensity moments at a single space-time point x are given by (6.22) in which $n=m$, and $x_1=x_2=\ldots=x_{2n}=x$. For a deep RPS, for which (6.26) is obeyed, the required functional derivative is evaluated in [6.72] with the result

$$\langle I^n\rangle = n!Y_n\left(\frac{\bar{N}\langle|R|^2\rangle}{1!},\frac{\bar{N}\langle|R|^4\rangle}{2!},\ldots,\frac{\bar{N}\langle|R|^{2n}\rangle}{n!}\right) \quad . \tag{6.76}$$

Here the Bell polynomials, $Y_n$, are defined as [6.213]

$$Y_n(x_1,x_2,\ldots,x_n) = \sum_{\{\nu_k\}}\frac{n!}{\nu_1!\nu_2!\ldots\nu_n!}\left(\frac{x_1}{1!}\right)^{\nu_1}\left(\frac{x_2}{2!}\right)^{\nu_2}\ldots\left(\frac{x_n}{n!}\right)^{\nu_n} \tag{6.77}$$

with $n=\nu_1+\nu_2+\ldots+\nu_n$ and the sum over all sets of numbers $\{\nu_k\}$ being solutions in nonnegative integers of the equation $\nu_1+2\nu_2+\ldots+n\nu_n=n$. The shorthand notation $\langle|R|^{2n}\rangle$ denotes an average over s and indicates the integration with the weight $n_0(\underline{\rho})$ over the illuminated area of the screen. Instances of (6.76) are

$$\begin{aligned}
\langle I\rangle &= \bar{N}\langle|R|^2\rangle\\
\langle I^2\rangle &= \bar{N}\langle|R|^4\rangle + 2\bar{N}^2\langle|R|^2\rangle^2\\
\langle I^3\rangle &= \bar{N}\langle|R|^6\rangle + 9\bar{N}^2\langle|R|^4\rangle\langle|R|^2\rangle + 6\bar{N}^3\langle|R|^2\rangle^3\\
\langle I^4\rangle &= \bar{N}\langle|R|^8\rangle + 16\bar{N}^2\langle|R|^6\rangle\langle|R|^2\rangle + 18\bar{N}^2\langle|R|^4\rangle^2\\
&\quad + 72\bar{N}^3\langle|R|^4\rangle\langle|R|^2\rangle^2 + 24\bar{N}^4\langle|R|^2\rangle^4 \quad .
\end{aligned} \tag{6.78}$$

In each of the above equations, it is the term with the highest power of $\bar{N}$ which is pertinent to Gaussian-field statistics, where $\langle I^n\rangle = n!\langle I\rangle^n$. The moments given by (6.76) describe a general situation indicated by (6.13) of an arbitrary amplitude of the incident beam, scattered by a screen with not necessarily uniform density of scatterers, $n_0(\underline{\rho})$, and a random shape factor s. Some particular cases treated by other authors are now easily recovered. In the works of JAKEMAN [6.214], and CHEN et al. [6.215-217], the moments were calculated for a fixed number N of the scattering centers, so that a comparison with our formulas requires the averaging over the Poisson distribution. The results of JAKEMAN correspond to uniform distribution of $u_0$ and $n_0$. Hence, in this case, $\langle |R|^{2n}\rangle = \langle s^{2n}\rangle$. On the other hand, in [6.215], the shape factor is constant, which corresponds to $\langle |R|^{2n}\rangle = s^{2n}$. The distribution and moments derived by PUSEY et al. [6.50,218,219] are obtained from (6.78) after identifying $\langle |R|^{2n}\rangle$ with an appropriate moment of the shape factor.

As shown in [6.72], the probability density function of the intensity can be found on the basis of (6.76) for the moments. Here we outline a different approach which also enables us to derive a joint probability density at a number of space-time points.

If the light is detected at L discrete points $x_1,\ldots,x_L$, we let $\zeta(x) = \sum \zeta_\ell \delta(x-x_\ell)$, which converts the characteristic functional (6.20) into a characteristic function of L complex variates $V_\ell$, $\ell = 1,\ldots,L$. The Fourier inversion formula yields then the field probability distribution $p(V_1,\ldots,V_L)$ and the intensity distribution at L points can, in principle, be expressed as

$$P(I_1,\ldots,I_L) = \int \delta(I_1-|V_1|^2)\ldots\delta(I_L-|V_L|^2)p(V_1,\ldots,V_L)d^2V_1\ldots d^2V_L \quad . \tag{6.79}$$

For $L = 1$, this gives

$$P(I) = \frac{1}{2}\exp(-\bar{N})\int_0^\infty \exp\left\{\bar{N}\langle J_0(u|R|)\rangle_{s,n_0}\right\}J_0(u\sqrt{I})udu \quad , \tag{6.80}$$

where $J_0$ is the Bessel function of zeroth order.

Equation (6.80) is the desired result for the probability density function of the intensity. If s is regarded as nonrandom, and the illumination is uniform, R is simply a constant and (6.80) reduces to

$$P(I) = \frac{1}{2}\int_0^\infty \exp\left\{\bar{N}[J_0(u|R|)-1]\right\}J_0(u\sqrt{I})udu \quad . \tag{6.81}$$

For large values of $\bar{N}$, (6.81) reproduces the exponential distribution corresponding to the Gaussian statistics of the scattered light, as can be seen by employing the original argument of RAYLEIGH [6.220]. To this end, we note that unless $u|R|$ is small, the factor $\exp\{\bar{N}[J_0(u|R|)-1]\}$ in (6.81) diminishes rapidly as $\bar{N}$ increases, since $\exp[J_0(u|R|)-1]$ is less than unity for any finite u. Thus, when $\bar{N}$ is very large, the important part of the range of integration corresponds to a small u for which the exponential integral in (6.81) may be replaced by $\exp(-\bar{N}u^2|R|^2/4)$. The leading term of the asymptotic formula is therefore

$$P(I) = (\bar{N}|R|^2)^{-1} \exp(-I/\bar{N}|R|^2) \quad . \tag{6.82}$$

Equation (6.81) can also be expressed as an average, weighted with the Poisson distribution, $\bar{N}^N \exp(-\bar{N})/N!$, of the familiar Kluyver-Rayleigh distribution describing a two-dimensional random walk with equal steps [6.219,221,222]. In fact, (6.81) can be written as

$$P(I) = \sum_{N=0}^{\infty} \frac{\bar{N}^N}{N!} \exp(-\bar{N}) P_N(I) \tag{6.83}$$

with

$$P_N(I) = \frac{1}{2} \int_0^{\infty} [J_0(u|R|)]^N J_0(u\sqrt{I}) u du \quad . \tag{6.84}$$

It is worthwhile to stress that one encounters (6.83) in quite a different context. If the first factor in the integrand of (6.84) is approximated by $\exp(-N|R|^2u^2/4)$, the integration can be carried out explicitly yielding

$$P(I) = \sum_{N=0}^{\infty} \exp(-\bar{N}) \frac{\bar{N}^N}{N!} \frac{\exp(-I/N|R|^2)}{N|R|^2} \quad . \tag{6.85}$$

This can be recognized as a formula for the total power density of the wave passing through an irregular diffracting medium [6.223,224]. In that case N is the number of times the wave has been scattered, and it is weighted with the Poisson factor having the mean value $\bar{N}$.

By employing the same procedure as we did to derive (6.80), the two-fold intensity distribution can be found. We merely point out that in the limit $\bar{N}\to\infty$, von LAUE's formula [6.45,75,79] is thus recovered.

### 6.4.2 Examples

Similarly to (6.76) for the moments, (6.80) embodies a number of situations depending on the conditions of illumination and the distributions of s and $n_0$. As the first example, let us consider the case of

Variable Density of Scatterers

To this end we assume that

$$n_0(\underline{\rho}) = \bar{n}_0 + n_{01}(\underline{\rho}) \quad . \tag{6.86}$$

Here $\bar{n}_0$ is thought of as representing a dense uniform background of small scattering grains, whereas $n_{01}(\underline{\rho})$ characterizes the distribution of large scatterers whose surface irregularities are of the order of the size of the illuminated region. These ideas are developed in detail in the studies devoted to statistical properties of a sum of speckles [6.55,225,226]. Accordingly, we write

$$\begin{aligned} &\bar{N}\int \bar{n}_0 \, d^2\rho \simeq \bar{N} \gg 1 \quad , \\ &\bar{N}\int n_{01}(\bar{\rho}) \, d^2\rho \approx 1 \quad . \end{aligned} \tag{6.87}$$

When, according to (6.86), the exponential in (6.80) is written as a product of two factors, and the factor containing $n_{01}$ is expanded in powers of $[J_0(u|R|)-1]$, then asymptotically for large $\bar{N}$, one obtains

$$P(I) = \frac{1}{\bar{N}|R|^2} \exp\left(-\frac{I+|R|^2}{\bar{N}|R|^2}\right) I_0\left(\frac{2|R|\sqrt{I}}{\bar{N}|R|^2}\right) \quad . \tag{6.88}$$

Thus, in this special case, the distribution function is given by the Rice-Nakagami law. It is interesting to observe that when (6.88) is extrapolated into the region of small $\bar{N}$, it becomes virtually identical with the log-normal distribution in which a variance parameter tends to zero. Explicit asymptotic expressions are given by STROHBEHN et al. [6.227]. The significance of the log-normal law in the context of the RPS was discovered by BLUEMEL et al. [6.228], who postulated this kind of distribution to fit their photocount data.

Uniform Illumination

The probability density of the reduced variable $i = I/|R|^2$ is given by (6.81), with $|R| = 1$, in accordance with the result of SCHAEFER and PUSEY [6.67]. Figure 6.5 shows the plot of P(i) obtained by numerical integration of (6.81).

Gaussian Spatial Profile of the Illuminating Beam

If the incident beam has the form

$$u_0(\underline{\rho}) = I_0^{1/2} \exp(-\rho^2/2a^2) \tag{6.89}$$

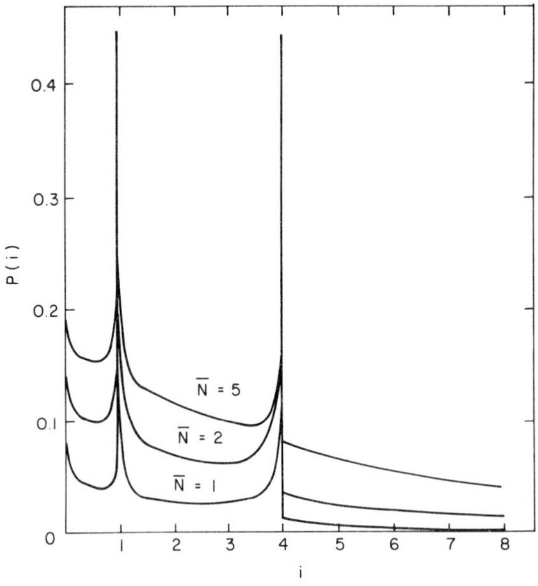


Fig. 6.5. Probability density function of the reduced variable $i = I/|R|^2$ for a few values of $\bar{N}$. Uniform illumination

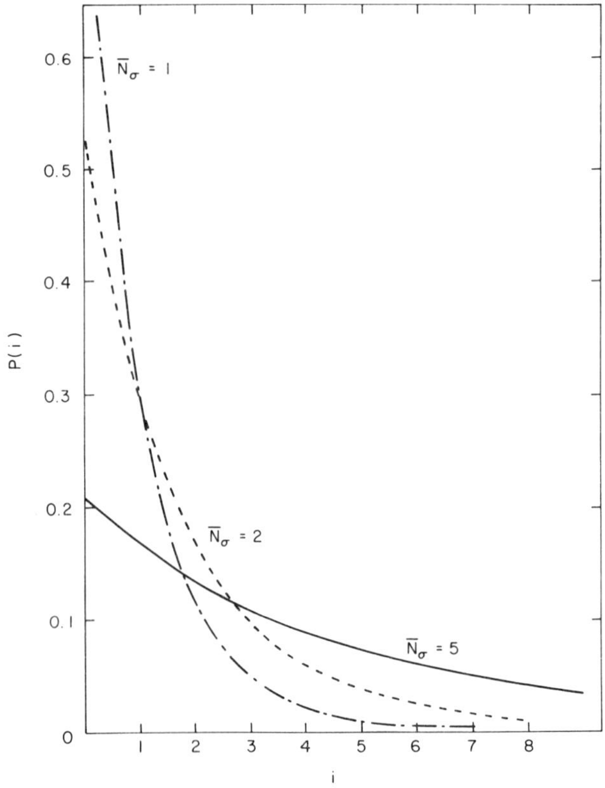


Fig. 6.6. Probability density function of the reduced variable $i = I/s^2I_0$ for a few values of $\bar{N}_\sigma$. Gaussian spatial profile of the illuminating beam

and $n_0 = \text{const}$, for nonrandom s we get

$$<|R|^{2n}> = \bar{N}_\sigma\ s^{2n}\ I_0^n/n \quad , \tag{6.90}$$

where $\bar{N}_\sigma = \bar{N} n_0 \pi a^2$ is the total number of scattering centers within the beam radius a. As can be seen from Fig.6.6, the probability density function of the reduced variable $i = I/s^2 I_0$ has no singularities occurring for uniform distribution. Equation (6.90) applies formally to the case of uniform illumination but random shape factor s arising from the Gaussian statistics of the gradient of the phase of the emerging light [6.19, 214]. Figure 6.6 then represents the probability density of the intensity observed at a given scattering angle.

## K-Distributions

Although P(I) cannot be evaluated analytically for arbitrary distribution of s, a class of model distributions has been proposed which leads to an agreement with experiment for a deep phase screen in the far-field region [6.50,176,229]. The modified Bessel or K-distribution

$$p(s) = \frac{2\beta}{\Gamma(1+\nu)} \left(\frac{\beta s}{2}\right)^{\nu+1} K_\nu(\beta s) \quad , \tag{6.91}$$

where $\nu$ and $\beta$ are constants dependent on the viewing position and incident intensity, results in an equation having the generic form (6.83) with

$$P_N(I) = \frac{\beta^{Q+1}\ I^{1/2(Q-1)}\ K_{Q-1}(\beta\sqrt{I})}{2^Q \Gamma(Q)} \quad , \tag{6.92}$$

where $Q = N(1+\nu)$.

Equation (6.92) lends itself to an experimental verification when the photocount distribution, which turns out to be the W-distribution [6.50], corresponding to (6.92) is derived.

### 6.4.3 Applications

Non-Gaussian statistics has proven a valuable source of information in light-scattering systems adequately described by the deep RPS model. The effect of dynamic scattering in liquid crystals was used to investigate the parameters of the screen, such as the rms phase deviation and the phase correlation length [6.24,230,231]. An analysis of the intensity probability distribution, as function of the spot size and correlation length of scattering inhomogeneities, was reported by SCUDIERI and BERTOLOTTI [6.232]. The work of WIENER-AVNEAR [6.233], where the nematic liquid crystal fluctuations were observed through the temporal changes in the magnified speckle picture, shows that these fluctuations result mostly from the thermal reorientation effect.

In the context of optical propagation through turbulence, the laboratory measurements of higher-order statistics were performed by mixing the convective air flow and the surrounding cooler air [6.17,176]. In order to fit the factorial moments of the photocount distribution, the K-distributions of Section 6.4.2 appear to be most appropriate.

Quasielastic light-scattering studies from colloidal solutions indicate non-Gaussian behavior at low densities associated with occupation number fluctuations. In homodyne detection, these fluctuations appear as an excess background with the characteristic time equal to the time required for a particle to pass through the scattering volume [6.218,234,235]. On the other hand, the characteristic time of the interference fluctuations is the time required for a typical particle to move the wavelength of light. The interference fluctuations, dominated by the Doppler fluctuation time, were used to determine the mobility of biological specimens [6.236-239]. The starting point of the analysis is the formula expressing the intensity correlation as

$$\langle I(0)I(t)\rangle = \langle \delta N(0)\delta N(t)\rangle + \langle N\rangle^2[1+|F(\underline{k},t)|^2] \quad , \tag{6.93}$$

which generalizes (6.78). Here $\delta N(t) = N(t) - \overline{N}$, and $F(\underline{k},t)$ is the phase autocorrelation function, $F(\underline{k},t) = \exp(-dk^2t)$, characterized by the diffusion constant d. Equation (6.93) is valid when the correlation time of the number fluctuations is much larger than the time over which $F(\underline{k},t)$ decays [6.234].

The predictions of the theory were tested using three strains of the bacterium *Escherichia coli* with different persistence lengths [6.240]. Through measurement of the correlation function of the scattered light intensity, it was possible to extract the values of the persistence length and the swimming speed.

## 6.5 Concluding Remarks

With reference to the inverse source problem, the far-field data can yield only statistical information about the scattering phase screen. The complete description of the screen as a random process in two dimensions, however attractive it may seem, is far beyond our reach at the present time. What can be achieved is the deduction of some low-order statistical features of a weak phase screen, such as the variance or the autocorrelation function of the surface profile. Examples of the input field characteristics are provided by (6.51) and (6.52). Remarkably enough, conclusions about the motion of the screen can already be drawn from the amplitude correlation function.

In the case of a deep RPS, the contrast-enhancing non-Gaussian fluctuations are of primary importance from the information-content point of view. By photocount techniques, the predictions of the theory can, in principle, be verified to include statistical moments of any desired order. The K-distributions, introduced in Section 6.4.2, determine the probability distribution of the diffraction shape factor, as well as higher-order moments. Establishing a link between the coherence properties of the scattered field and the higher-order statistical features of the scattering screen remains probably the most promising open problem of the theory.

*Acknowledgments.* The author wishes to thank Drs. H.H. Arsenault, H.P. Baltes, J. Chrostowski, C. Delisle, H. Fujii, J.C. Leader and B. Steinle for stimulating conversations and correspondence.

## References

6.1 H.G. Booker, J.A. Ratcliffe, D.H. Shin: Phil. Trans. Roy. Soc. (London) *242*, 579 (1950)
6.2 A. Hewish: Proc. Roy. Soc. London *209*, 81 (1951)
6.3 J.A. Ratcliffe: Rep. Prog. Phys. *19*, 188 (1956)
6.4 L.T. Little, A. Hewish: Mon. Not. R. Astron. Soc. *138*, 393 (1968)
6.5 E. Jakeman, E.R. Pike, P.N. Pusey: Nature *263*, 215 (1976)
6.6 L.S. Taylor: J. Math. Phys. *13*, 590 (1972)
6.7 W. Martienssen, E. Spiller: Am. J. Phys. *32*, 919 (1964)
6.8 W. Martienssen, E. Spiller: Phys. Rev. Lett. *16*, 531 (1966)
6.9 F.T. Arecchi: Phys. Rev. Lett. *15*, 912 (1965)
6.10 C. Deutsch, P.N. Keating: J. Appl. Phys. *40*, 4049 (1969)
6.11 G.H. Heilmeier, L.A. Zanoni, L.A. Barton: Proc. IEEE *56*, 1162 (1968)
6.12 V.H. Rumsey: Radio Sci. *10*, 107 (1975)
6.13 M. Marians: Radio Sci. *10*, 115 (1975)
6.14 Y. Furuhama: Radio Sci. *10*, 1037 (1975)
6.15 A.M. Prokhorov, F.V. Bunkin, K.S. Gochelashvily, V.I. Shishov: Proc. IEEE *63*, 790 (1975)
6.16 R.L. Fante: Proc. IEEE *63*, 1669 (1975)
6.17 E. Jakeman, J.G. McWhirter, G. Parry, P.N. Pusey: Paper presented at the *Topical Meeting on Optical Propagation Through Turbulence, Rain and Fog* (Golem Press, Boulder, Colo. 1977) WC1-1
6.18 A. Rocca, F. Roddier, J. Vernin: J. Opt. Soc. Am. *64*, 1000 (1974)
6.19 E. Jakeman, P.N. Pusey: J. Phys. A. *8*, 369 (1975)
6.20 J.C. Leader: J. Opt. Soc. Am. *66*, 536 (1976)
6.21 E. Jakeman, J.G. McWhirter, P.N. Pusey: J. Opt. Soc. Am. *66*, 1175 (1976)
6.22 G. Parry: "Speckle Patterns in Partially Coherent Light", in *Laser Speckle and Related Phenomena*, ed. by J.C. Dainty, Topics in Applied Physics, Vol.9 (Springer, Berlin, Heidelberg, New York 1975) p.76
6.23 H.M. Pedersen: Opt. Commun. *12*, 156 (1974)
6.24 E. Jakeman, P.N. Pusey: J. Phys. A. *6*, L88 (1975)
6.25 L. Mandel, E. Wolf: Rev. Mod. Phys. *37*, 231 (1965)
6.26 G.S. Agarwal: Opt. Commun. *14*, 161 (1975)
6.27 E. Kretschmann: Opt. Commun. *10*, 353 (1974)
6.28 E. Kretschmann, E. Kröger: J. Opt. Soc. Am. *65*, 150 (1975)
6.29 F. Toigo, A. Marvin, V. Celli, N.R. Hill: Phys. Rev. B *15*, 5618 (1977)
6.30 M.S. Longuet-Higgins: Phil. Trans. Roy. Soc. (London) *249*, 321 (1957)

6.31 R.P. Mercier: Proc. Cambr. Phil. Soc. *58*, 382 (1962)
6.32 M.V. Berry: Phil. Trans. Roy. Soc. London *273*, 611 (1973)
6.33 L. Allen, D.G.C. Jones: Phys. Lett. *7*, 321 (1963)
6.34 M. Rousseau, D. Canals-Frau: C. R. Acad. Sci. Ser. B *269*, 514 (1969)
6.35 P. Beckmann, A. Spizzino: *The Scattering of Electromagnetic Waves from Rough Surfaces* (Pergamon/McMillan, London, New York 1963)
6.36 P. Beckmann: "Scattering of Light by Rough Surfaces", in *Progress in Optics*, ed. by E. Wolf, Vol.6 (North-Holland, Amsterdam 1967) p.55
6.37 P.J. Chandley, W.T. Welford: Opt. Quant. Elect. *7*, 393 (1975)
6.38 H.M. Pedersen: Opt. Acta *22*, 523 (1975)
6.39 A.S. Marathay, L. Heiko, J.L. Zuckermann: Appl. Opt. *9*, 2470 (1970)
6.40 L.H. Enloe: Bell Syst. Tech. J. *46*, 1479 (1976)
6.41 L.E. Estes, L.M. Narducci, R.A. Tuft: J. Opt. Soc. Am. *61*, 1301 (1971)
6.42 J.W. Strutt (Lord Rayleigh): Phil. Mag. *10*, 73 (1880)
6.43 J.W. Strutt (Lord Rayleigh): *Theory of Sound* (Dover, New York 1945) Sect.42
6.44 M. von Laue: Sitzungsber. Akad. Wiss. (Berlin) *44*, 1144 (1914)
6.45 M. von Laue: Mitt. Physik. Ges. (Zürich) *18*, 90 (1916)
6.46 M. von Laue: Verhandl. Deut. Phys. Ges. *19*, 19 (1917)
6.47 A. Sommerfeld: *Optics* (Academic Press, New York 1967) p.191
6.48 M. Born, E. Wolf: *Principles of Optics* (Pergamon, London 1970)
6.49 B. Crosignani, P. DiPorto, M. Bertolotti: *Statistical Properties of Scattered Light* (Academic Press, New York 1975)
6.50 P.N. Pusey: "Statistical Properties of Scattered Radiation" in *Photon Correlation Spectroscopy and Velocimetry*, ed. by H.Z. Cummins, E.R. Pike (Plenum Press, New York 1977) p.45
6.51 L.I. Goldfisher: J. Opt. Soc. Am. *55*, 247 (1965)
6.52 J.C. Leader: J. Opt. Soc. Am. *68*, 175-185 (1978)
6.53 J.C. Leader: J. Opt. Soc. Am. *67*, 1091 (1977)
6.54 V.F. Sudakov: Opt. Spektrosk. *40*, 604 (1976)
6.55 D.L. Fried: J. Opt. Soc. Am. *66*, 1150 (1976)
6.56 N. George: J. Opt. Soc. Am. *66*, 1182 (1976)
6.57 J.C. Erdmann, R.I. Gellert: J. Opt. Soc. Am. *66*, 1194 (1976)
6.58 P. Croce, L. Prod'homme: Nouv. Rev. Opt. *7*, 121 (1976)
6.59 M.J. Beran, G.B. Parent: *Theory of Partial Coherence* (Prentice-Hall, Englewood Cliffs 1964)
6.60 J. Peřina: Phys. Lett. *12*, 194 (1964)
6.61 R.S. Ingarden: Fortschr. Phys. *13*, 755 (1965)
6.62 E.F. Keller: Phys. Rev. *139*, B202 (1965)
6.63 A. Zardecki: J. Math. Phys. *11*, 244 (1970)
6.64 A. Zardecki: J. Phys. A. *7*, 2198 (1974)
6.65 S. Chandrasekhar: Rev. Mod. Phys. *15*, 1 (1943)
6.66 M. Rousseau: J. Opt. Soc. Am. *61*, 1307 (1971)
6.67 D.W. Schaefer, P.N. Pusey: "Statistics of Light Scattered by Non-Gaussian Fluctuations", in *Coherence and Quantum Optics*, ed. by L. Mandel, E. Wolf (Plenum Press, New York 1973) p.839
6.68 F.A. Haight: *Handbook of the Poisson Distribution* (Wiley and Sons, New York 1967) Chaps.3-4
6.69 B. Picinbono, C. Bendjaballah, J. Pouget: J. Math. Phys. *11*, 2166 (1970)
6.70 A. Blanc-Lapierre, R. Fortet: *Théorie des Fonctions Aléatoires* (Masson, Paris 1953) Chap.5
6.71 Y. Ichioka: J. Opt. Soc. Am. *64*, 919 (1974)
6.72 A. Zardecki, C. Delisle: Opt. Acta *24*, 241 (1977)
6.73 R.J. Glauber: "Optical Coherence and Photon Statistics", in *Quantum Optics and Electronics*, ed. by C. De Witt, A. Blandin, C. Cohen-Tannoudji (Gordon and Breach, New York 1965) p.65
6.74 J.R. Klauder, E.C.G. Sudarshan: *Fundamentals of Quantum Optics* (Benjamin, New York 1968)
6.75 J.C. Dainty: "The Statistics of Speckle Patterns", in *Progress in Optics*, ed. by E. Wolf, Vol.14 (North-Holland, Amsterdam 1976) p.1
6.76 H.Z. Cummins: "Laser Light Scattering Spectroscopy", in *Quantum Optics*, ed. by R.J. Glauber (Academic Press, New York 1969) p.259
6.77 A. Zardecki, C. Delisle: Can. J. Phys. *51*, 1017 (1973)

6.78 T.S. McKechnie: Opt. Commun. *13*, 35 (1975)
6.79 W. Goodman: "Statistical Properties of Laser Speckle Patterns", in *Laser Speckle and Related Phenomena*, ed. by J.C. Dainty, Topics in Applied Physics, Vol.9 (Springer, Berlin, Heidelberg, New York 1975) p.9
6.80 L. Mandel: Proc. Phys. Soc. London *74*, 233 (1959)
6.81 D. Gabor: "Light and Information" in *Progress in Optics*, ed. by E. Wolf, Vol.I (North-Holland, Amsterdam 1961) p.107
6.82 A. Zardecki, C. Delisle, J. Bures: Opt. Commun. *5*, 298 (1972)
6.83 J. Peřina, L. Mišta: Opt. Acta *21*, 329 (1974)
6.84 C.W. Helstrom: Proc. Phys. Soc. London *83*, 777 (1964)
6.85 C.W. Helstrom: J. Opt. Soc. Am. *60*, 521 (1970)
6.86 C.W. Helstrom: J. Opt. Soc. Am. *67*, 833 (1977)
6.87 F. Gori, G. Guatori: J. Opt. Soc. Am. *64*, 453 (1974)
6.88 F. Gori, S. Paolucci, L. Ronchi: J. Opt. Soc. Am. *65*, 495 (1975)
6.89 V. Blazek: Opt. Commun. *11*, 144 (1974)
6.90 J. Bures: J. Opt. Soc. Am. *64*, 1598 (1974)
6.91 M. Elbaum, P. Diament: Appl. Opt. *15*, 2268 (1976)
6.92 B.E.A. Saleh: J. Opt. Soc. Am. *67*, 71 (1977)
6.93 A. Zardecki, C. Delisle, J. Bures: "Volume of Coherence", in *Coherence and Quantum Optics*, ed. by L. Mandel, E. Wolf (Plenum Press, New York 1973) p.259
6.94 R. Barakat: Opt. Acta *20*, 729 (1973)
6.95 H.H. Arsenault, G. April: J. Opt. Soc. Am. *66*, 1160 (1976)
6.96 L. Mandel: J. Opt. Soc. Am. *51*, 1342 (1961)
6.97 L. Mandel, E. Wolf: Phys. Rev. *124*, 1696 (1961)
6.98 L. Mandel, E. Wolf: J. Opt. Soc. Am. *66*, 529 (1976)
6.99 L. Mandel: Proc. Phys. Soc. London *81*, 1104 (1963)
6.100 A.A. Scribot: Opt. Commun. *13*, 81 (1975)
6.101 Y. Rezette: Opt. Commun. *16*, 86 (1976)
6.102 P. Beckmann: *The Depolarization of Electromagnetic Waves* (Golem Press, Boulder, Colo. 1968)
6.103 J.C. Leader: J. Appl. Phys. *42*, 4808 (1971)
6.104 J.C. Leader, W.A.J. Dalton: J. Appl. Phys. *43*, 3080 (1972)
6.105 G.J. Wilhelmi, J.W. Rouse, Jr., A.J. Blanchard: J. Opt. Soc. Am. *65*, 1036 (1975)
6.106 K. Gåsvik: Opt. Commun. *22*, 61 (1977)
6.107 G.J. Wilhelmi, J.C. Leader, W.A.J. Dalton: Appl. Opt. *15*, 1837 (1976)
6.108 J.W. Goodman: Proc. IEEE *53*, 1688 (1965)
6.109 J.W. Goodman: Opt. Commun. *13*, 244 (1975)
6.110 J.C. Dainty: Opt. Acta *18*, 327 (1971)
6.111 A.A. Scribot: Opt. Commun. *11*, 238 (1974)
6.112 T.S. McKechnie: In *Recent Advances in Optical Physics* ed. B. Havelka, J. Blabla (Palacky University, Olomouc, Society of Czechoslovak Mathematicians and Physicists, Prague 1976) p.97
6.113 M.G. Miller, A.M. Schneiderman, P.F. Kellen: J. Opt. Soc. Am. *65*, 779 (1975)
6.114 S. Wadaka, T. Sato: J. Opt. Soc. Am. *66*, 145 (1976)
6.115 G. Parry: Opt. Acta *21*, 763 (1974)
6.116 G. Parry: Opt. Commun. *12*, 75 (1974)
6.117 G. Parry: Opt. Quant. Elect. *7*, 311 (1975)
6.118 H.M. Pedersen: Opt. Acta *22*, 15 (1975)
6.119 N. George, A. Jain: Appl. Opt. *12*, 1202 (1973)
6.120 E. Jakeman, E.R. Pike, G. Parry, B. Saleh: Opt. Commun. *19*, 359 (1976)
6.121 W.T. Welford: Opt. Quant. Elect. *9*, 269 (1977)
6.122 H.P. Baltes, B. Steinle: Nuovo Cimento *18*, 318 (1977)
6.123 H.P. Baltes, B. Steinle: Nuovo Cimento B. *41*, 428 (1977)
6.124 H.P. Baltes, B. Steinle, G. Antes: "Radiometric and Correlation Properties of Bounded Planar Sources", in *Proc. 4th Rochester Conference on Coherence and Quantum Optics*, June 8-10, 1977, ed. L. Mandel, E. Wolf (Plenum Press, New York 1978)
6.125 J.A. Holzer, C.C. Sung: J. Appl. Phys. *47*, 3363 (1976)
6.126 J.C. Leader, A.K. Fung: J. Appl. Phys. *48*, 1736 (1977)
6.127 K. Nagata, T. Umebara: Jpn. J. Appl. Phys. *12*, 694 (1973)

6.128 K. Nagata, T. Umebara, J. Nishiwaki: Jpn. J. Appl. Phys. *12*, 1693 (1973)
6.129 O.K. Taganov, A.S. Toporets: Opt. Spektrosk. *40*, 423 (1976)
6.130 P.J. Chandley: Opt. Quant. Elect. *8*, 323 (1976)
6.131 P.J. Chandley: Opt. Quant. Elect. *8*, 329 (1976)
6.132 N. Takai: Opt. Commun. *14*, 24 (1975)
6.133 J.D. Litster: "Liquid Crystals", in *Photon Correlation and Light Beating Spectroscopy*, ed. H.Z. Cummins, E.R. Pike (Plenum Press, New York 1974) p.475
6.134 J.A. Castellano: Opt. Laser Technol. *7*, 259 (1975)
6.135 S. Chandrasekhar: Rep. Prog. Phys. *39*, 613 (1976)
6.136 R. Bartolino, M. Bertolotti, F. Scudieri, D. Sette: Appl. Opt. *12*, 2917 (1973)
6.137 F. Scudieri, M. Bertolotti, R. Bartolino: Appl. Opt. *13*, 181 (1974)
6.138 M. Bertolotti, M. Carnevale, B. Daino, D. Sette: Appl. Opt. *9*, 962 (1970)
6.139 M. Bertolotti, P. DiPorto , B. Crosignani: Phys. Rev. A *5*, 396 (1972)
6.140 A.P. Chaikovskii: Opt. Spektrosk. *40*, 76 (1976)
6.141 M. Bertolotti, F. Scudieri, S. Verginelli: Appl. Opt. *15*, 1842 (1976)
6.142 S. Lowenthal, H.H. Arsenault: J. Opt. Soc. Am. *60*, 1487 (1970)
6.143 H.H. Arsenault: J. Opt. Soc. Am. *61*, 1425 (1971)
6.144 J.W. Goodman: J. Opt. Soc. Am. *66*, 1145 (1976)
6.145 R.A. Sprague: Appl. Opt. *11*, 2811 (1972)
6.146 W.T. Welford: Opt. Quant. Elect. *7*, 413 (1975)
6.147 H.M. Pedersen: Opt. Commun. *16*, 63 (1973)
6.148 H.M. Pedersen: J. Opt. Soc. Am. *66*, 1204 (1976)
6.149 H. Fujii, T. Asakura: Opt. Commun. *11*, 35 (1974)
6.150 H. Fujii, T. Asakura: Opt. Commun. *12*, 32 (1974)
6.151 J. Ohtsubo, H. Fujii, T. Asakura: Jpn. J. Appl. Phys. Suppl. *14-1*, 293 (1975)
6.152 H. Fujii, T. Asakura: Nouv. Rev. Opt. *6*, 5 (1975)
6.153 H. Fujii, T. Asakura, Y. Shindo: Opt. Commun. *16*, 68 (1976)
6.154 H. Fujii, T. Asakura, Y. Shindo: J. Opt. Soc. Am. *66*, 1217 (1976)
6.155 H. Fujii, J. Uozumi, T. Asakura: J. Opt. Soc. Am. *66*, 1222 (1976)
6.156 H. Fujii, T. Asakura: Appl. Opt. *16*, 180 (1977)
6.157 H. Fujii, T. Asakura: Opt. Commun. *21*, 80 (1977)
6.158 H. Fujii, J. Lit: Opt. Commun. *22*, 231 (1977)
6.159 L.H. Tanner: Opt. Laser Techn. *8*, 113 (1976)
6.160 J.M. Bennet: Appl. Opt. *15*, 2705 (1976)
6.161 J.C. Richmond, J.J. Hsia: J. Res. Nat. Bur. Stand. *80*A, 207 (1976)
6.162 J.W. Goodman: Opt. Commun. *14*, 324 (1975)
6.163 J. Ohtsubo, T. Asakura: Appl. Opt. *16*, 1742 (1977)
6.164 T. Fukaya, J. Tsujiuchi: Nouv. Rev. Opt. *6*, 317 (1975)
6.165 E. Jakeman, W.T. Welford: Opt. Commun. *21*, 72 (1977)
6.166 G. Ross: Opt. Acta *15*, 451 (1968)
6.167 G. Ross: Opt. Acta *16*, 611 (1969)
6.168 G. Ross: Phil. Trans. Roy. Soc. (London) *268*, 177 (1970)
6.169 J. Uozumi, H. Fujii, T. Asakura: J. Opt. Soc. Am. *67*, 808 (1977)
6.170 G. Kortüm: *Reflectance Spectroscopy* (Springer, Berlin, Heidelberg, New York 1969)
6.171 E.P. Lavin: *Specular Reflection* (American Elsevier, New York 1971)
6.172 G.J. Brownsey, J.W. Eldridge, D.A. Jarvis, G. Ross, I. Sanders: J. Phys. E. *9*, 654 (1976)
6.173 E. Jakeman, J.G. McWhirter: J. Phys. A. *9*, 785 (1976)
6.174 G. Parry, P.N. Pusey, E. Jakeman, J.G. McWhirter: Opt. Commun. *22*, 195 (1977)
6.175 E. Jakeman, J.G. McWhirter: J. Phys. A. *10*, 1599 (1977)
6.176 G. Parry, P.N. Pusey, E. Jakeman, J.G. McWhirter: "The Statistical and Correlation Properties of Light Scattered by a Random Phase Screen", in *Proc. 4th Rochester Conference on Coherence and Quantum Optics*, ed. L. Mandel, E. Wolf (Plenum Press, New York 1978)
6.177 E. Archbold, J.M. Burch, A.E. Ennos: Opt. Acta *17*, 883 (1970)
6.178 E. Archbold, A.E. Ennos: Opt. Acta *19*, 253 (1972)
6.179 A.W. Lohman, G.P. Weigelt: Opt. Commun. *14*, 252 (1975)
6.180 G.P. Weigelt: Opt. Commun. *19*, 222 (1976)
6.181 A.W. Lohman, G.P. Weigelt: J. Opt. Soc. Am. *66*, 1271 (1976)
6.182 A.W. Lohman, G.P. Weigelt: Opt. Commun. *20*, 50 (1977)
6.183 M. Françon, P. Koulev, M. May: Opt. Commun. *12*, 63 (1974)

6.184 M. Françon, P. Koulev, M. May: Opt. Commun. *13*, 138 (1975)
6.185 M. May, M. Françon: J. Opt. Soc. Am. *66*, 1275 (1976)
6.186 J.A. Mendez, M.L. Roblin: Opt. Commun. *15*, 226 (1975)
6.187 V.N. Thinh, S. Tanaka: Opt. Commun. *20*, 367 (1977)
6.188 S. Komatsu, I. Yamaguchi, H. Sato: Opt. Commun. *18*, 314 (1976)
6.189 L. Celaye, J.M. Jonathan, S. Mallick: Opt. Commun. *18*, 496 (1976)
6.190 D.A. Gregory: Opt. Laser Techn. *8*, 201 (1976)
6.191 D.A. Gregory: Opt. Commun. *20*, 1 (1977)
6.192 Y. Dzialowski, M. May, R. Shaw: Opt. Commun. *21*, 282 (1977)
6.193 D.B. Barker, M.E. Fourney: Opt. Lett. *1*, 135 (1977)
6.194 R. Jones: Opt. Laser Techn. *8*, 215 (1976)
6.195 R.P. Khetan, F.P. Chiang: Appl. Opt. *15*, 2205 (1976)
6.196 L. Ek, N.E. Nolin: Opt. Commun. *2*, 184 (1970)
6.197 H.J. Tiziani: Appl. Opt. *11*, 2911 (1972)
6.198 A.E. Ennos: "Speckle Interferometry", in *Laser Speckle and Related Phenomena*, ed. by J.C. Dainty, Topics in Applied Physics, Vol.9 (Springer, Berlin, Heidelberg, New York 1975) p.203
6.199 P.N. Pusey: J. Phys. D *9*, 1399 (1976)
6.200 H.H. Arsenault, S. Lowenthal: Opt. Commun. *1*, 451 (1970)
6.201 S. Lowenthal, D. Joyeux, H.H. Arsenault: Opt. Commun. *2*, 184 (1970)
6.202 B. Crosignani, B. Daino, P. Di Porto: J. Appl. Phys. *42*, 399 (1971)
6.203 N. Takai: Jpn. J. Appl. Phys. *13*, 2025 (1974)
6.204 V.I. Voronin, A.S. Dunaev, R.D. Mukhamedyarov: Sov. J. Opt. Technol. *42*, 293 (1975)
6.205 L.A. Östlund, K. Biedermann: Appl. Opt. *16*, 685 (1977)
6.206 B.E.A. Saleh: Appl. Opt. *14*, 2344 (1975)
6.207 E. Jakeman: J. Phys. A *8*, 123 (1975)
6.208 I. Yamaguchi, S. Komatsu: Opt. Acta *24*, 705 (1977)
6.209 E. Ingelstam, S.I. Ragnarsson: Vision Res. *12*, 411 (1972)
6.210 D.A. Palmer: Vision Res. *16*, 436 (1976)
6.211 L. Ronchi, A. Fontana: Opt. Acta *22*, 243 (1975)
6.212 L.H. Tanner: Appl. Opt. *13*, 2026 (1974)
6.213 J. Riordan: *An Introduction to Combinatorial Analysis* (Wiley and Sons, New York 1958) Chap.2
6.214 E. Jakeman: "Photon Correlation", in *Photon Correlation and Light Beating Spectroscopy*, ed. by H.Z. Cummins, E.R. Pike (Plenum Press, New York 1974) p.75
6.215 S.H. Chen, P. Tartaglia: Opt. Commun. *6*, 119 (1972)
6.216 P. Tartaglia, S.H. Chen: Opt. Commun. *7*, 379 (1973)
6.217 S.H. Chen, P. Tartaglia, P.N. Pusey: J. Phys. A *6*, 490 (1973)
6.218 D.W. Schaefer, P.N. Pusey: Phys. Rev. Lett. *29*, 843 (1972)
6.219 P.N. Pusey, D.W. Schaefer, D.E. Koppel: J. Phys. A *7*, 530 (1974)
6.220 J.W. Strutt (Lord Rayleigh): Phil. Mag. (6) *37*, 321 (1919)
6.221 R. Barakat: Opt. Acta *21*, 903 (1974)
6.222 R. Barakat, J. Blake: Phys. Rev. A *13*, 1122 (1976)
6.223 J.A. Feyer: Proc. Roy. Soc. London A*220*, 455 (1953)
6.224 E.N. Bromley: Proc. Roy. Soc. London A*225*, 515 (1954)
6.225 J.K. Jao, M. Elbaum: J. Opt. Soc. Am. *67*, 1266 (1977)
6.226 J. Ohtsubo, T. Asakura: Opt. Lett. *1*, 98 (1977)
6.227 J.W. Strohbehn, T.-I. Wang, J.P. Speck: Radio Sci. *10*, 59 (1975)
6.228 V. Bluemel, L.M. Narducci, R.A. Tuft: J. Opt. Soc. Am. *62*, 1309 (1972)
6.229 E. Jakeman, P.N. Pusey: IEEE Trans. AP-*24*, 806 (1976)
6.230 E. Jakeman, P.N. Pusey: Phys. Lett. A *44*, 456 (1973)
6.231 P.N. Pusey, E. Jakeman: J. Phys. A *8*, 392 (1975)
6.232 F. Scudieri, M. Bertolotti: J. Opt. Soc. Am. *64*, 776 (1974)
6.233 E. Wiener-Avnear: Appl. Phys. Lett. *29*, 635 (1976)
6.234 D.W. Schaefer, B.J. Berne: Phys. Rev. Lett. *28*, 475 (1972)
6.235 F.D. Carlson: Annu. Rev. Biophys. Bioeng. *4*, 243 (1975)
6.236 R. Nossal, S.H. Chen, C.C. Lai: Opt. Commun. *4*, 35 (1971)
6.237 R. Nossal, S.H. Chen: Opt. Commun. *5*, 117 (1972)
6.238 H. Shimizu, G. Matsumoto: Opt. Commun. *16*, 197 (1976)
6.239 R.V. Mustacich, B.R. Ware: Phys. Rev. Lett. *33*, 617 (1974)
6.240 D.W. Schaefer, B.J. Berne: Biophys. J. *15*, 785 (1975)

# Additional References with Titles


## Chapter 1

Frieden, B.R.: "Image Enhancement and Restoration", in *Picture Processing and Digital Filtering*, ed. by T.S. Huang, Topics in Applied Physics, Vol.6 (Springer, Berlin, Heidelberg, New York 1967) pp.177-248

Frieden, B.R., Wells, D.C.: Restoring with maximum entropy. III. Poisson sources and backgrounds. J. Opt. Soc. Am. *68*, 93-103 (1978)

Kermisch, D.: A deterministic analysis of the maximum entropy image restoration method and of some related methods. J. Opt. Soc. Am. *67*, 1154-1159 (1977)

Bertero, M., De Mol, C., Viano, G.A.: Restoration of optical objects using regularization. Opt. Lett. *3*, 51-53 (1978)

Bertero, M., De Mol, C., Viano, G.A.: On the problems of object restoration and image extrapolation in optics. Submitted to J. Math. Phys.

Casasent, D. (ed.): *Optical Data Processing*, Topics in Applied Physics, Vol.23 (Springer, Berlin, Heidelberg, New York 1978)

Bates, R.H.T., Lewitt, R.M., McDonnell, M.J., Milner, M.O., Peters, T.M.: Practical image processing. Phys. Technol. *9*, 101-107 (1978)

Deepak, A. (ed.): *Inversion Methods in Atmospheric Remote Sounding* (Academic Press, New York 1977)

Ishimaru, A.: *Wave Propagation and Scattering in Random Media* (Academic Press, New York 1978)

Goedecke, G.H.: Radiative transfer in closely packed media. J. Opt. Soc. Am. *67*, 1339-1346 (1977)

## Chapter 2

Ross, G., Fiddy, M.A., Nieto-Vesperinas, M., Wheeler, M.W.L.: A solution to the phase problem based on the theory of entire functions. Optik *49*, 71-80 (1977)

Montgomery, W.D.: Phase retrieval and the polarization identity. Opt. Lett. *2*, 120-121 (1978)

Fienup, J.R.: Reconstruction of an object from the modulus of its Fourier transform. Opt. Lett. *3*, 27-29 (1978)

Psaltis, D., Casasent, D.: Phase determination of an amplitude modulated complex wavefront. Appl. Opt. *17*, 1136-1140 (1978)

Ohtsuka, Y.: Proposal for the determination of the complex degree of spatial coherence. Opt. Lett. *1*, 133-134 (1977)

## Chapter 3

Bleistein, N., Bojarski, N.N.: Recently developed formulations of the inverse problem in acoustics and electromagnetics. University of Denver, Colo., Rt.No. MS-R-7501 (1974)

Nishimura, M., Psaltis, D., Caimi, F., Casasent, D.: Implementation of the inverse radon transform by optical convolution. Opt. Commun. *25*, 301-304 (1978)
McWhirter, J.G., Pike, E.R.: On the numerical inversion of the Laplace transform and similar Fredholm integral equations of the first kind. J. Phys. A (1978) to be published
Hoenders, B.J., Baltes, H.P.: The scalar theory of nonradiating partially coherent sources. Submitted to J. Phys. A
Hoenders, B.J., Baltes, H.P.: On the existence of nonradiating stochastic current distributions. Submitted to J. Phys. A

Chapter 4

Walther, A.: Gabor's theorem and energy transfer through lenses. J. Opt. Soc. Am. *57*, 639-644 (1967)
Ross, G.: A high-resolution light-scattering method for studying the fine-scale structure of polymers. J. Phys. D *6*, 1537-1549 (1973)

Chapter 5

Eckhardt, W.: Anisotropy and temporal coherence of blackbody radiation. Phys. Rev. A *17*, 1093-1099 (1978)
Carpenter, D.J., Pask, C.: The angular spectrum approach to diffraction of partially coherent light. Opt. Acta *24*, 939-948 (1977)
Bastiaans, M.J.: The Wigner distribution function applied to optical signals and systems. Opt. Commun. *25*, 26-30 (1978)
Ueha, S., Oshima, S., Tsujiuchi, J.: Image reconstruction by using the propagation law of mutual intensity. Opt. Commun. *18*, 488-491 (1976)
Leader, J.C.: Far-zone range-criteria for quasihomogeneous partially coherent sources. Preprint
Carver, K.R.: Radiometric recognition of coherence. Radio Science *12*, 371-379 (1977)
Tatarskii, V.I.: "Locally Homogeneous Fields with Smoothly Varying Mean Characteristics", in *The Effects of the Turbulent Atmosphere on Wave Propagation* (Israel Program for Scientific Translations, Jerusalem 1971) §7
Wolf, E.: Coherence and radiometry. J. Opt. Soc. Am. *68*, 6-17 (1978)
Carter, W.H.: Radiant intensity from inhomogeneous sources and the concept of averaged cross-spectral density. Opt. Commun. *26*, 1-4 (1978)
Boivin, A., Deckers, C.: Un nouveau théorème d'eschantillonnage pour le degré de cohérence complexe en vue de la reconstitution d'une source lumineuse isotrope. Opt. Commun. *26*, 144-147 (1978)
Baltes, H.P., Hoenders, B.J.: $K_\nu$ correlations and micro-area models in diffuse scattering. To be presented at the Intern- Conf. on Lasers, Orlando, Fla., Dec.11-15 (1978)

Chapter 6

Lorentz, H.A.: On the change in intensity in the diffraction pattern of a large number of irregulary arranged holes or particles. Versl. K. Adak. Wet. Amsterdam *26*, 1120 (1918)
Ross, G.: Light scattering in amorphous media. Experimental method. Opt. Acta *16*, 95-109 (1969)
Ross, G.: Light scattering in amorphous media. The object wave and its coherence. Opt. Acta *25*, 57-66 (1978)
Ross, G., Fiddy, M.A.: The speckle effect: a reappraisal. Opt. Acta *25*, 205-217 (1978)

Jakeman, E., McWhirter, J.G.: Correlation function dependence of the scintillation behind a deep random phase screen. J. Phys. A *10*, 1599-1643 (1977)
Parry, G., Pusey, P.N., Jakeman, E., McWhirter, J.G.: Focusing by a random phase screen. Opt. Commun. *22*, 195-201 (1977)
Jakeman, E., Pusey, P.N.: Significance of K distributions in scattering experiments. Phys. Rev. Lett. *40*, 546-550 (1978)
Chrostowski, J., Zardecki, A.: The effect of occupation number fluctuations on partially developed Speckle patterns. Opt. Commun. *26*, 27-30 (1978)
Ohtsubo, J., Asakura, T.: Measurement of surface roughness properties using speckle patterns with non-Gaussian statistics. Opt. Commun. *25*, 315-319 (1978)
Sung, C.C., Eberhardt, W.D.: Explanation of the experimental results of light backscattered from a very rough surface. J. Opt. Soc. Am. *68*, 323-328 (1978)
Carter, W.H., Bertolotti, M.: An analysis of the far-field coherence and radiant intensity of light scattered from liquid crystals. J. Opt. Soc. Am. *68*, 329-333 (1978)
Leader, J.C.: Intensity fluctuations resulting from partially coherent light propagating through atmospheric turbulence. Preprint
de Wolf, D.A.: Waves in random media: Weak scattering reconsidered. J. Opt. Soc. Am. *68*, 475-479 (1978)
Fried, D.L.: Propagation of the mutual coherence function for an infinite plane wave through a turbid medium. Opt. Lett. *1*, 104-106 (1977)
Lee, M.H.: Variance and covariance of irradiance of a finite beam in extremely strong turbulence. J. Opt. Soc. Am. *68*, 167-169 (1978)

# Subject Index

# Springer Series in Optical Sciences

Editor: D.L. MacAdam

*A Selection*

Volume 2: R. Beck, W. Englisch, K. Gürs

## Table of Laser Lines in Gases and Vapors

2nd revised and enlarged edition. 1978.
IX, 202 pages
ISBN 3-540-08603-X

The book is a compilation of all laser lines in gases and vapors. It contains about 5.000 lines, which are arranged in two different tables by the active medium and by wavelength. The operating conditions and the relevant references are given. Potential readers are all scientists and engineers, working in the field of laser research, development and applications.

Volume 5: M. Young

## Optics and Lasers

An Engineering Physics Approach
1977. 122 figures, 4 tables. XIV, 207 pages
ISBN 3-540-08126-7

This is an up-to-date text that surveys applied optics, includng classical ray and wave optics, lasers, holography and coherent optics. An attempt was made to treat each topic in sufficient depth to give it considerable engineering value, while keeping it as free of unnecessary mathematical detail as possible. Admittedly interesting phenomena such as the formation of the rainbow or the precise determination of the speed of light have been skipped for abvious reasons. The text should be useful to practicing engineers, physicists, or graduate students. The level of any of the material is appropriate to an introductory course in optics. Problems are designed not only to help increase the reader's understanding but also, sometimes, to derive a useful result. (When the result of a problem is of general importance, the result is always quoted).

Volume 6: B. Saleh

## Photoelectron Statistics

With Applications to Spectroscopy and Optical Communication
1978. 85 figures, 8 tables. XV, 441 pages
ISBN 3-540-08295-6

This book presents a concise unified treatment of the statistical properties of light fields of various origins and the statistics of their corresponding photoelectrons. It outlines the role played by photoelectron fluctuations in the areas of optical communication and photon-correlation spectroscopy.

Volume 10: R.H. Kingston

## Detection of Optical and Infrared Radiation

1978. 39 figures, 2 tables. VIII, 140 pages
ISBN 3-540-08617-X

This book covers the fundamentals of optical and infrared detection including the properties of the radiation, the ideal photodetector, coherent or heterodyne detection, noise processes in vacuum photodiodes, photomultipliers, photoconductors, semiconductor photodiodes, and avalanche photodiodes. Also included is a treatment of the ideal thermal detector, the bolometer and the pyroelectric detector, followed by an analysis of the laser as a low-noise amplifier preceding the detector. Following a chapter on the effects of the atmosphere on detector performance, detection and error probabilities are discussed for both Gaussian noise and photon counting limits. The final chapter applies the theory to several applications including radiometry, spectroscopy, radar, and astronomical interferometry. This unified treatment of detectors and the detection process is the first such comprehensive text in the past decade.

Springer-Verlag
Berlin
Heidelberg
New York